Janna Krummenacker

Investigations on the high cycle fatigue strength of short glass fiber reinforced polyamide 66

FSC
www.fsc.org
MIX
Papier aus verantwortungsvollen Quellen
Paper from responsible sources
FSC® C105338

Bibliografische Information Der Deutschen Bibliothek: Die Deutsche Bibliothek verzeichnet diese Publikation in der Deutschen Nationalbibliografie; detaillierte bibliografische Daten sind im Internet über dnb.dnb.de abrufbar.

Bibliographic information published by Die Deutsche Bibliothek: Die Deutsche Bibliothek lists this publication in the Deutsche Nationalbibliografie; detailed bibliographic data is available in the Internet at dnb.dnb.de.

Herstellung und Verlag: BoD – Books on Demand, Norderstedt

ISBN: 978-3-7562-5626-6

Investigations on the high cycle fatigue strength of short glass fiber reinforced polyamide 66

Vom Fachbereich Maschinenbau und Verfahrenstechnik
der Technischen Universität Kaiserslautern
zur Erlangung des akademischen Grades

Doktor-Ingenieur (Dr.-Ing.)
genehmigte
Dissertation

von
Frau
Dipl.-Ing. Janna Krummenacker
aus Norden

Tag der mündlichen Prüfung: 02.06.2022

Prüfungsvorsitzender: Prof. Dr.-Ing. Ulf Breuer
1. Berichterstatter: Prof. Dr.-Ing. Joachim Hausmann
2. Berichterstatter: Prof. Valter Carvelli, PhD

D 386

Vorwort

Die vorliegende Arbeit entstand im Rahmen meiner Tätigkeit als wissenschaftliche Mitarbeiterin an der Leibniz-Institut für Verbundwerkstoffe GmbH im Programmbereich Bauteilentwicklung im Kompetenzfeld Ermüdung und Lebensdaueranalyse.

Mein besonderer Dank gilt meinem Doktorvater Herrn Prof. Dr.-Ing. Joachim Hausmann für die wissenschaftliche Betreuung der Arbeit, die stetige Unterstützung und das mir entgegengebrachte Vertrauen. Ebenso danke ich Herrn Prof. Valter Carvelli für die Übernahme der Berichterstattung und die ehrliche Diskussion meiner Arbeit. Bedanken möchte ich mich außerdem bei Herrn Prof. Dr.-Ing. Ulf Breuer für die Übernahme des Prüfungsvorsitzes.

Darüber hinaus bedanke ich mich bei allen Mitarbeitenden des Instituts für die zahlreiche Unterstützung und die freundschaftliche Arbeitsatmosphäre. Besonders hervorheben möchte ich Christian Becker, Matthias Bendler, Tobias Donhauser, Esha, Nora Feiden, Stefan Gabriel, Werner Gölzer, Francis Gonzalez-Ramirez, Valentine Kessler, Andreas Klingler, Regina Köhne, Konstantin Mehl, Nicole Motsch-Eichmann, Thomas Pfaff, Jan Rehra, Stefan Schmidt, Ralph Schneider und Julia Vogtmann. Auch bei meinen ehemaligen Kollegen Yves Becker, Marcel Bücker, Hermann Giertzsch, Benedikt Hannemann, Torsten Heydt, Benjamin Kelkel, Thomas Rief, Florian Rieger und Florian Schimmer möchte ich mich hiermit bedanken.

Ein besonderer Dank geht an meine beiden Bürokollegen in dieser Zeit Constantin Bauer und Andreas Baumann für die zahlreichen fachlichen Diskussionen, die große Hilfsbereitschaft und nicht zuletzt die freundschaftlichen Gespräche.

Mein Dank geht auch an Prof. Dr.-Ing. Michael Magin für die hilfreiche Diskussion meiner Arbeit, die genau zum richtigen Zeitpunkt kam. Ebenso bedanken möchte ich mich bei Florian Regitz, David Heepen und Jonas Lehnhoff für die tatkräftige Unterstützung als Hilfswissenschaftler.

Ein großes Dankeschön geht an Marius Orfgen, der mir zahllose Male geholfen hat, einen anderen Blickwinkel auf meine Arbeiten zu erlangen und aus der ein oder anderen Sackgasse herauszukommen.

Ein besonderer Dank geht an meine Freundinnen, die mich nicht nur bei dieser Arbeit unterstützt haben, sondern mich schon fast mein ganzes Leben begleiten: Anna, Ariane, Caro, Eva, Evelyn, Lisa und Sarah. Angefangen vom Bewerbungsschreiben, über die Motivation in schwierigen Phasen bis zum Korrekturlesen - auf euch kann ich immer zählen. Vielen Dank!

Zu guter Letzt möchte ich mich bei meiner Familie für ihre uneingeschränkte Unterstützung bedanken. Vielen Dank an meinen Freund Mathias - mit niemandem sonst hätte ich mich auf das Abenteuer „Promotion, Familie gründen und Haus renovieren“ eingelassen. Die Aussicht mit ihm und unserem Sohn Paul Zeit zu verbringen, war stets die größte Motivation auch in anstrengenden Phasen nicht zu prokrastinieren. Ein besonderer Dank geht an meine Eltern, die mir stets jeden Freiraum gegeben haben und deren Rückhalt und Vertrauen mich durchs Leben trägt.

Kaiserslautern, Juni 2022

Janna Krummenacker

Table of contents

Abstract

The growth of composite materials is closely linked to the demands of the transportation industry for efficient and resource-saving solutions. The beginnings of modern lightweight construction lie in the aviation industry, which constantly strives to reduce weight in order to achieve greater travel distances and reduce fuel consumption. For this purpose, continuous fiber-reinforced plastics are used, that are characterized in particular by their outstanding specific mechanical properties and therefore meet the high requirements for load-bearing capacity in aviation. In the automotive sector, the focus is similar and increasingly trends towards weight-optimized solutions in order to meet the growing requirements to reduce CO_2 emissions. Substituting metallic components with composite materials can be a solution for these challenges. However, the requirements for the materials used in the automotive sector are different from those in the aerospace sector. In contrast to the rather low production volumes in the aerospace industry, vehicle components are manufactured in large quantities and therefore require short cycle times and cost-efficient production.

One group of materials that meet these requirements are short-fiber-reinforced thermoplastics (SFRTs). They are characterized in particular by their low-cost manufacturing, which can be achieved through the injection molding process and therefore enables the production of components in high volumes. Glass fibers are mainly used for reinforcement, with the fiber length limited to 1 mm due to the injection molding process. Despite this short fiber length, the strength and stiffness of the pure polymer can be significantly increased with short fiber reinforcement. However, the reinforcement effect is strongly influenced by the fiber orientation resulting from the injection molding process. Layers with different fiber orientation are formed, which are accompanied by a pronounced anisotropic material behavior. Due to the thermoplastic matrix, the mechanical properties of SFRT additionally depend on environmental conditions such as humidity and temperature. Consequently, characterizing the material and thereby covering all influencing factors requires effortful and lengthy testing. Especially when determining fatigue properties for a detailed service life analysis, long testing times combined with high costs must be taken into account. In this case, a quasi-static strength analysis is usually performed for the dimensioning of components, which is often also used with empirical reduction factors for the fatigue analysis.

Considering the above mentioned context, the present work investigates the behavior of short glass fiber-reinforced polyamide 66 in the range of very high cycle fatigue (> 10^6 load cycles) and addresses the question as to whether a fatigue limit exists for the aforementioned material or not. For this particular purpose, various experimental methods are used, which all have in common, that they generate large amounts of data requiring automated processing. The experiments are carried out with test specimens longitudinally and transversely to the injection molding direction in order to observe a possible influence of the fiber orientation on the material's fatigue behavior.

In the first part of this experimental work, the stiffness and the hysteresis data are studied during fatigue tests with different maximum stress levels. Subsequently, the stiffness degradation and the dissipative energy are parameterized to correlate this data with the applied maximum stresses. The data analysis method identifies characteristic stress levels at which fatigue behavior changes. The changes occur in three fatigue life ranges of low cycle, high cycle and very high cycle fatigue. The research shows that a cyclic load with 10^5 cycles is sufficient to estimate the three mentioned fatigue ranges.

In the second part of this experimental work, residual strength tests with acoustic emission (AE) analysis are performed on the cyclically preloaded specimens. AE uses sensors to detect acoustic signals generated by crack initiation and crack growth under mechanical load. Accordingly, in the residual strength tests, only the damage that has not already occurred under cyclic preloading can be recorded. Based on the "acoustic fingerprint", characteristic stress levels were identified at which a change in damage behavior occurred. As long as this "acoustic fingerprint" differs from that of a non-preloaded specimen, it can be concluded that damage was initiated under the applied cyclic load.

Finally, a digital twin was generated to investigate the underlying micromechanical mechanisms at the experimentally identified characteristic stress levels. X-ray microscope scans of the specimens were imported into the commercial software GeoDict®. This voxel-based software uses numerically efficient Fast Fourier Transforms (FFT) to analyze the microstructure and simulate the experiments on the real 3D microstructure. The simulations show that the fiber-matrix interface significantly influences the damage behavior in the very high cycle fatigue range. By analyzing the matrix plasticization rate, the stress levels associated with high cycle fatigue can be estimated. Fiber fractures, on the other hand, are only relevant in the low cycle fatigue range.

Thus, both the experimental methods and the simulation of a digital twin have been proven suitable to estimate the different fatigue ranges in a time-efficient manner. In addition, the analyses of stiffness degradation in fatigue tests and acoustic emissions in residual strength tests indicate that damage occurs at a loading above stresses, which correspond to a fatigue life of 10^{11} cycles. This in turn implies that no fatigue limit exists before 10^{11} load cycles for the short-fiber-reinforced polyamide 66 under investigation.

Kurzfassung

Das Aufkommen von Verbundwerkstoffen ist eng gekoppelt mit den Forderungen des Transportwesens nach effizienten und ressourcenschonenden Transportmitteln. Die Anfänge des modernen Leichtbaus liegen in der Luftfahrtbranche, die stetige Gewichtseinsparungen anstrebt, um größere Reichweiten zu erzielen und den Kraftstoffverbrauch zu verringern. Dabei kommen endlosfaserverstärkte Kunststoffe zum Einsatz, die sich insbesondere durch ihre herausragenden spezifischen mechanischen Eigenschaften auszeichnen und somit die hohen Ansprüche an die Tragfähigkeit in der Luftfahrt erfüllen. Auch in der Automobilbranche liegt der Fokus immer stärker auf gewichtsoptimierten Lösungen, um die zunehmenden Anforderungen an eine Reduktion der CO_2-Emissionen zu erfüllen. Hierauf kann die Substitution von metallischen Komponenten durch Verbundwerkstoffe eine Antwort sein. Jedoch liegen hier andere Anforderungen an die verwendeten Werkstoffe als in der Luftfahrtbranche vor. Im Gegensatz zu den vergleichsweise geringen Produktionsmengen in der Luftfahrtindustrie müssen Fahrzeugkomponenten in kurzen Taktzeiten und kosteneffizient produziert werden können.

Eine Werkstoffgruppe, die diese Anforderungen erfüllt, sind kurzfaserverstärkte Thermoplaste, die sich insbesondere durch ihre kostengünstige Herstellung im Spritzgussverfahren auszeichnen, was eine Produktion von Bauteilen in hohen Stückzahlen ermöglicht. Als Verstärkungsfasern kommen hier hauptsächlich Glasfasern zum Einsatz, die durch den Spritzgussprozess auf Faserlängen unterhalb von 1 mm begrenzt sind. Trotz dieser geringen Faserlänge kann mit einer Kurzfaserverstärkung eine deutliche Steigerung der reinen Polymer-Festigkeit und -Steifigkeit erzielt werden. Diese Verstärkungswirkung wird jedoch stark von der Faserorientierung beeinflusst, die sich aus dem Spritzgussverfahren ergibt. Es bilden sich Schichten mit unterschiedlicher Faserorientierung, womit ein ausgeprägt anisotropes Materialverhalten einhergeht. Aufgrund der thermoplastischen Matrix sind die mechanischen Eigenschaften kurzfaserverstärkter Thermoplaste außerdem abhängig von Umgebungsbedingungen wie Feuchte und Temperatur. Daher ist eine alle Einflussfaktoren umfassende Materialcharakterisierung mit einem hohen Prüfaufwand verbunden. Insbesondere bei der Ermittlung von Ermüdungseigenschaften müssen somit lange Prüfzeiten verbunden mit hohen Kosten einkalkuliert werden, bis eine detaillierte Lebensdaueranalyse möglich ist. Daher erfolgt die Auslegung von Strukturbauteilen in der Regel mittels eines quasi-statischen

Festigkeitsnachweises, der häufig nach Berücksichtigung empirischer Reduktionsfaktoren auch als Lebensdauernachweis herangezogen wird.

Vor diesem Hintergrund beschäftigt sich die vorliegende Arbeit mit dem Verhalten von kurzglasfaserverstärktem Polyamid 66 im Bereich sehr hoher Lastwechselzahlen (> 10^6) und untersucht die Frage, ob eine Dauerfestigkeit für den genannten Werkstoff existiert. Hierfür kommen verschiedene experimentelle Methoden zum Einsatz, die alle gemein haben, dass große Datenmengen erzeugt werden und automatisiert verarbeitet werden müssen. Die Versuche werden für Probekörper in der und quer zur Spritzgießrichtung durchgeführt, um einen möglichen Einfluss der Faserorientierung auf die Vorgehensweise zu erfassen.

Im ersten Teil der experimentellen Arbeiten werden die Steifigkeits- und die Hysteresedaten während Ermüdungsversuchen mit unterschiedlichen maximalen Spannungen untersucht. Anschließend findet die Parametrisierung der Steifigkeitsdegradation sowie der dissipativen Energie statt, um diese Daten mit den aufgebrachten maximalen Spannungen in Bezug zu setzen. Mit dieser Form der Datenanalyse werden verschiedene charakteristische Spannungsniveaus identifiziert, an denen sich das Ermüdungsverhalten ändert. Diese Änderungen treten im Bereich der drei Lebensdauerbereiche *low cycle*, *high cycle* und *very high cycle fatigue* auf. Darüber hinaus zeigen die Untersuchungen, dass mit einer zyklischen Belastung von 10^5 Lastwechseln eine Abschätzung der drei genannten Lebensdauerbereiche möglich ist.

In dem zweiten Teil der experimentellen Arbeiten werden an den zyklisch vorbelasteten Proben Restfestigkeitsversuche mit Schallemissionsanalyse durchgeführt. Bei der Schallemissionsanalyse detektieren Schallsensoren akustische Signale, die unter mechanischer Last durch Rissentstehung und Risswachstum entstehen. In den Restfestigkeitsversuchen können demnach nur die Schädigungen aufgezeichnet werden, die nicht schon vorher unter zyklischer Last entstanden sind. Auf Grundlage des „akustischen Fingerabdrucks" wurden charakteristische Spanungsniveaus identifiziert, bei denen ein Wechsel des Schädigungsverhaltens aufgetreten ist. Solange sich dieser „akustische Fingerabdruck" von dem einer unbelasteten Probe unterscheidet, kann die Schlussfolgerung gezogen werden, dass unter der applizierten zyklischen Last Schädigungen initiiert wurden.

Abschließend wurde ein digitaler Zwilling erzeugt, um die mikromechanischen

Mechanismen auf den experimentell identifizierten, charakteristischen Spannungsniveaus zu untersuchen. Dazu werden mit einem Röntgenmikroskop Scans der Probekörper erzeugt, die in die kommerzielle Software GeoDict® importiert werden. Die Software arbeitet voxelbasiert und greift auf numerisch effiziente Fast Fourier Transformationen zurück, um die Mikrostruktur zu analysieren und eine Simulation der Versuche an der realen 3D Mikrostruktur durchzuführen. Die durchführten Simulationen zeigen, dass das Faser-Matrix-Interface das Schädigungsverhalten im Bereich sehr hoher Lastwechselzahlen (*very high cycle fatigue*) maßgeblich beeinflusst. Über eine Analyse der Matrixplastifizierungsrate können die mit dem *high cycle fatigue* verbundenen Spannungsniveaus abgeschätzt werden. Faserbrüche werden hingegen erst im *low cycle fatigue* Bereich relevant.

Somit haben sich sowohl die experimentellen Methoden als auch die Simulation eines digitalen Zwillings als geeignet gezeigt, um die verschiedenen Ermüdungsbereiche zeiteffizient abzuschätzen. Außerdem konnte mit Hilfe der Steifigkeitsdegradationsanalyse und der akustischen Emissionsanalyse in Restfestigkeitsversuchen gezeigt werden, dass für das untersuchte kurzfaserverstärkte Polyamid 66 eine Dauerfestigkeit nicht vor 10^{11} Lastwechseln existiert.

Glossary

List of abbreviations

Abbreviation	Description
AE	acoustic emission
AI	artificial intelligence
AIC	Akaike information criterion
CT	computed tomography
DIC	digital image correlation
FFT	Fast Fourier Transform
FRP	fiber reinforced polymers
HCF	high cycle fatigue
LCF	low cycle fatigue
PA6.6	polyamide 6.6
RVE	representative volume elements
SFRT	short fiber reinforced thermoplastics
VHCF	very high cycle fatigue

Latin variables

a	-	orientation tensor, parameter of the S-N curve
A	-	asymmetric wave mode
b		parameter of the S-N curve
E	MPa	Young's modulus
E_{dyn}	N/mm	dynamic stiffness
f		frequency
F	-	material effort / failure index
F	N	force
G	MPa	shear modulus
n	-	sample
N	-	last sample, number of cycles
N_f		lifetime
p	-	orientation vector
P	-	probability
R	-	load ratio, signal
R_{UTS}	MPa	ultimate tensile strength

R^2		coefficient of determination
s	mm	displacement
S	-	symmetric wave mode, stress level
t	s	time
v	m/s	sound velocity
x	mm	parameter
y	mm	parameter
Y	MPa	Yield stress
x, y, z		spatial directions

Greek variables

α		first hardening parameter
β		second hardening parameter
θ	°	angle to discribe fiber orientation, angle to direction of injection molding
ε	%	strain
μ		micro-, mean (average)
σ	MPa/-	stress, standard deviation
τ	MPa	shear stress
ν	-	Poisson ratio
Φ	°	angle to discribe fiber orientation
Ψ	-	probability distribution

Indices

0	basic wave modes, start
1	spatial direction in the injection molding direction
2	spatial direction transverse to the injection molding direction
core	core layer
i, j	directional variables
m	mean
max	maximum
min	minimum
origin	origin of acoustic event
s	survival
shell	shell layer
t	tensile

Tsai-Hill	Tsai-Hill failure criterion
u	ultimate strength
UTS	ultimate tensile strength
w	window

Mathematical signs

e	Euler's number ≈ 2.72
$\log_{10}$	Logarithm with base 10
π	Archimedes's constant ≈ 3.14

1 Introduction

The widespread use of composites in the aerospace industry, where the focus is on lightweight design, has shown that composites are suitable for saving weight in the transportation sector. Due to their high performance, thermoset prepregs made of carbon fiber-reinforced epoxy resin are usually used, which can be layered with optimized fiber orientations to achieve maximum load-bearing capacity. The disadvantage of these thermoset composites is the high raw material and manufacturing costs, which, however, plays only a minor role in aerospace due to the low production volumes.

While lightweight construction and weight savings are also important in the automotive sector, the requirements for the materials used are different from those in the aerospace industry. Due to their high cost, the application of continuous fiber-reinforced composites in the automotive sector is mainly limited to luxury and sports vehicles. Moreover, many components do not require the high mechanical properties of these high-performance materials, which is why many metallic components in the automotive sector are being replaced by thermoplastic composites. The focus is on cost-effective base materials and manufacturing processes that are characterized by short cycle times in order to be able to produce large quantities.

This is where SFRTs come into the game. Their greatest advantage is their manufacturability in the injection molding process, which results in short cycle times, variable molding and high reproducibility with low quality scatter. Compared to continuous fiber-reinforced thermosets, however, their mechanical properties such as stiffness and strength are less favorable, which is why load-bearing components have so far mostly been made of continuous fiber-reinforced laminates. This is the reason why the majority of the research on the mechanical behavior of fiber-reinforced composites focuses on continuous fiber-reinforced composites.

With the increasing use of SFRT in load bearing components, it is becoming more and more important to investigate the mechanical behavior of these materials. The mechanical properties of components made of SFRT are influenced by numerous factors, which strongly depend on the process parameters during manufacturing. The injection molding process affects both the fiber length and the fiber orientation distribution, and causes several layers of different fiber orientation to form over the component thickness. This results in a pronounced anisotropic material behavior. The three parameters mentioned, such as fiber length, fiber orientation distribution and degree of anisotropy, have a decisive influence on the reinforcing effect of the fibers. The longer the fibers, the better the load transfer from the polymer into the fibers. A high fiber content combined with a uniform fiber distribution also leads to a higher reinforcing effect. Finally, the mechanical properties such as strength and stiffness are significantly higher parallel to the fiber direction than transverse to it.

In practice, the design of components is usually carried out by means of a quasi-static strength analysis, which is often also used as a service life analysis with empirical reduction factors. For continuous fiber-reinforced laminates, it has proven useful to limit the strain under cyclic loading to 0.3%-0.6%, thus assuming that this limit corresponds to the fatigue strength. TALREJA discusses the determination of this strain limit considering the matrix toughness and fiber structure [1]. The inaccuracy of this empirical value, which is commonly used in aviation industry but lacks of scientific basis, is confronted with the long test duration required to characterize fatigue behavior, especially in the high cycle fatigue range. For SFRT, this experimental effort becomes even more pronounced when considering the numerous factors that affect the mechanical properties as a result of the complex microstructure.

Therefore, this work investigates experimental methods that allow predicting the behavior of SFRT in the range of very high numbers of tensile-tensile load cycles in a time-efficient manner. The very high cycle fatigue range is understood here to apply to cyclic loading exceeding 10^6 load cycles. Furthermore, the existence of a so-called endurance limit for SFRT is investigated, i.e. a fatigue stress below which no failure occurs regardless of the number of load cycles. For this purpose, the focus is on taking advantage of the possibilities offered by today's computers and modern analysis methods for evaluating large amounts of data.

The fatigue behavior of SFRT is often estimated using phenomenological approaches, which correlate the increasing damage under cyclic loading with different parameters such as stiffness degradation [2,3] or the dissipative energy [4–6]. These parameters have the advantage of being easily accessible without further testing equipment. At the same time, the large amounts of data generated during a fatigue test series cannot be evaluated manually. Therefore, in this work, the stiffness degradation and dissipative energy are analyzed with new methods to gain further insight into the fatigue behavior of SFRT in a time-efficient manner.

A second approach explored in this work, which also requires advanced evaluation methods for the large amounts of data, is acoustic emission analysis in a mechanical materials test. The operating principle of AE testing attributes the released acoustic emissions to damage mechanisms such as crack initiation and growth [7]. AE in materials testing has shown that it can successfully contribute to the characterization of damage progression under mechanical loading [8]. In the present work, AE was investigated during quasi-static tensile tests on previously non-preloaded specimens and on specimens preloaded by fatigue. This procedure aims to determine the presence of a critical stress level below which no detectable damage is caused by fatigue loading.

Finally, the findings obtained from the experimental data are examined by means of a micromechanical model. This model is created using the voxel-based software GeoDict® [9].

2 State of the art

The following section provides an overview of the constituents, properties and damage mechanisms of short glass fiber reinforced polyamide 6.6 (PA6.6), followed by an explanation of the test methods used in this work.

2.1 Short glass fiber reinforced thermoplastics

The matrix material investigated here, PA6.6, belongs to the group of thermoplastics. Unlike thermosetting polymers, thermoplastics bond through intermolecular forces and therefore melt when heated. This physical condition leads to several advantages, such as recyclability, weldability and the possibility to reshape the material after production. The molecular chains of PA6.6 are partially aligned which classifies them as semi-crystalline plastics, resulting in them having not only a melting point but also one or more glass transitions. Since the glass transition of PA6.6 occurs at room temperature, it behaves visco-elastically in most technical applications. This is the reason why the mechanical properties of SFRT depend significantly on the environmental conditions such as temperature and humidity [3,10,11].

Glass fibers are used to increase the strength and stiffness of the matrix material. To produce glass fibers, glass is melted, passed through spinnerets and then drawn out into a uniform and round cross-section. The typical diameter is between 5-24 µm. Rapid cooling creates an amorphous structure, giving the glass fibers isotropic properties. Compared to other reinforcing fibers, glass fibers stand out in particular for their low price. Glass fibers are divided into different categories according to their structure. In technical applications, the use of so-called E-glass fibers (**e**lectrical application) is predominant. E-glass has a Young's modulus of approx. 73 GPa and a tensile strength of approx. 2.6 GPa. Other common types of glass fibers are C-glass (**c**hemical) and S-glass (**s**trength) for applications requiring higher chemical resistance and higher tensile strength, respectively.

To ensure that the high strength of the fibers can be fully used, the load must be transferred from the matrix to the fibers via a so-called fiber-matrix-interface. The load transfer is significantly influenced by two factors: the quality of the interface and the fiber length. To achieve good adhesion, the fiber has to be sized before it is processed. In addition, the sizing protects the fibers from abrasion during further processing. Fiber lengths are commonly divided into three groups. Fibers with a length of less than 1 mm are referred to as short fibers. Fibers with a length between 1 mm and 50 mm belong to the group of so-called long fibers. With a length exceeding 50 mm, fibers are assigned to the category of continuous fibers. With increasing fiber length, their reinforcing effect increases as well. At the same time, short and long fibers can be used in cost-effective processing methods such as injection molding. The shorter the fibers, the easier it is for them to reach all areas of the mold, even with complex geometries.

To take advantage of the many benefits of injection molding, short fibers are exclusively combined with thermoplastic polymers. In the injection molding process, the glass fiber-reinforced thermoplastic is melted and injected into a mold. After it has cooled of, it falls out of the mold making space for the production of the next part. The main advantages of injection molding are the short cycle periods and the high level of automation, which in turn lead to low variations in material quality.

The operating parameters of the injection molding process affect the fiber length, fiber distribution, and fiber orientation and result in the formation of several layers over the component thickness with different fiber orientation. The material that is in contact with the mold walls cools off very quickly resulting in random fiber orientation in a very thin, outermost layer. This layer is referred to as skin layer. The flow velocity increases from the outside to the inside leading to shear stresses within the molten material. This forms the so-called shell layer, in which the fibers are mainly aligned in the flow direction. An extensional flow with a small velocity gradient in the center forms a core layer in which the fibers are mostly oriented transverse to the flow direction [12]. The thickness of the respective layers is influenced on the one hand by the viscosity of the melt and on the other hand by the component thickness.

Figure 2.1 shows X-ray images of a plate with a thickness of 2 mm and 3 mm and visualizes the above-explained structure with different fiber orientation in layers. The figure also illustrates that the degree of fiber orientation depends on the plate thickness. The ratio between the shell and the core layer is greater for the 2 mm than for the 3 mm thick plate, resulting in a stronger fiber orientation in the direction of injection molding for the 2 mm thick plate. The observation that the degree of anisotropy increases with decreasing thickness was also confirmed by DE MONTE, MOOSBRUGGER AND QUARESIMIN [13].

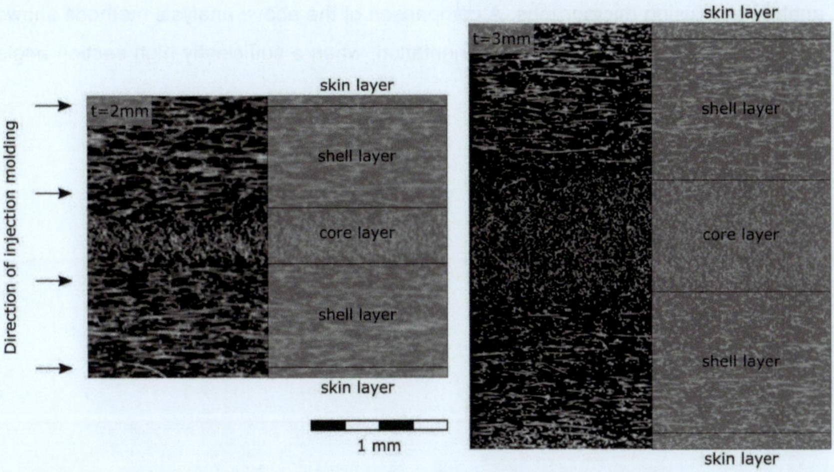

Figure 2.1: Layered structure with different fiber orientation for plates with a thickness of 2 mm and 3 mm.

Fiber orientation can be determined by analyzing microsections or by using 3D imaging techniques such as micro-computed tomography (µCT) or X-ray microtomography. The advantage of the analysis of microsections is the low-cost and the good availability of measuring equipment. However, sample preparation is time-consuming and fiber orientation calculation is complicated. For the fiber orientation analysis by 3D imaging techniques, samples are scanned with a resolution below the fiber diameter.

While the sample preparation effort is low, this method is associated with high equipment costs. The scans can be evaluated with commercial software either using classical image processing methods or artificial intelligence (AI). Most available tools use classical image processing methods that distinguish matrix and fibers based on their gray values. This evaluation method requires little computational power and averages fiber orientation for defined areas of the scan. With AI techniques, the individual voxels are assigned to either matrix or fiber material. This provides the user with additional information such as fiber length and fiber thickness distribution. However, it requires a multiple voxel mapping of the fibers resulting in a higher resolution of the scans and a higher computational effort.

Another advantage of fiber orientation analysis using 3D scans is that, depending on the relationship between specimen size and required resolution, the complete specimen can be analyzed and does not have to be destroyed, as is the case when analyzing by using micrographs. A comparison of the above analysis methods shows good agreement in predicting fiber orientation, when a sufficiently high section angle of the microsections is provided [14].

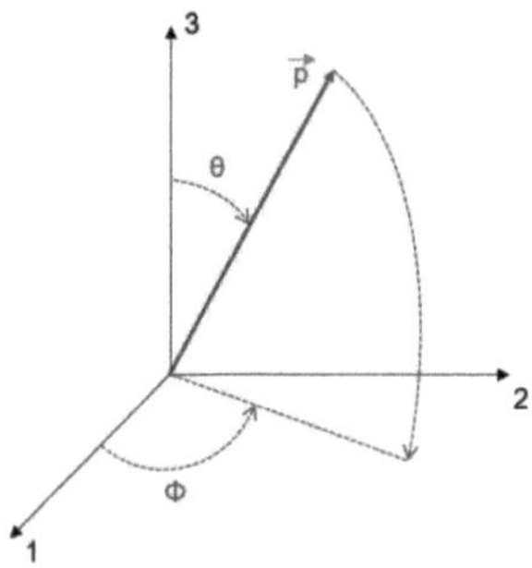

$$\vec{p} = \begin{pmatrix} \sin\theta \cdot \cos\Phi \\ \sin\theta \cdot \sin\Phi \\ \cos\theta \end{pmatrix}$$

Figure 2.2: Orientation vector $\boldsymbol{p}$.

A position and an orientation vector $\boldsymbol{p}$ can describe the orientation of a single fiber (see Figure 2.2). To capture the orientation of all fibers in an injection molded material, ADVANI AND TUCKER [15] used a probability distribution $\Psi(\boldsymbol{p})$, which defines the probability of finding a fiber at a given point with a given orientation. They introduced the so-called orientation tensor $\boldsymbol{a}_{ij}$, which is suitable for the numerical description of the fiber orientation distribution. Equation 2.1 shows the orientation tensor defined as the dyadic product of the orientation vector p weighted by the distribution function Ψ.

$$a_{ij} = \oint p_i \otimes p_j \Psi(p) dp \qquad 2.1$$

The orientation tensor is symmetric with a trace equal to one and it can be visualized by an ellipsoid where the principal axes direction and length correspond to the eigenvectors respectively to the eigenvalues of $\boldsymbol{a}_{ij}$. The longer the ellipsoid, the more pronounced the fiber orientation, or the rounder the ellipsoid, the more isotropically oriented the fibers. A principal axis transformation provides the three eigenvectors and eigenvalues. The two extreme fiber orientation distributions can be represented by a line and a sphere. The line illustrates a unidirectional fiber orientation. In this case, the first eigenvalue is 1 while the other two eigenvalues are 0. The sphere, on the other hand, corresponds to the visualization of an isotropic fiber orientation distribution where all three eigenvalues have the value 1/3.

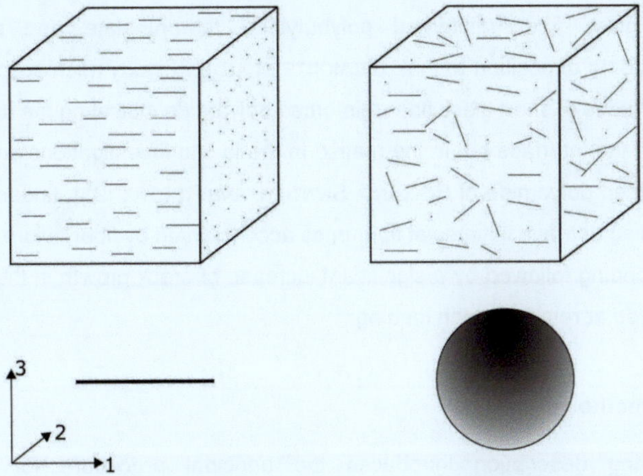

$$a_{ij} = \begin{bmatrix} 1 & 0 & 0 \\ 0 & 0 & 0 \\ 0 & 0 & 0 \end{bmatrix} \qquad a_{ij} = \begin{bmatrix} 1/3 & 0 & 0 \\ 0 & 1/3 & 0 \\ 0 & 0 & 1/3 \end{bmatrix}$$

Figure 2.3: Fiber orientation visualization with associated orientation tensors a_{ij}.

2.2 Damage behavior of SFRT

The fiber orientation has a major influence on the mechanical properties of SFRT. The mechanical properties and fatigue strengths in the direction of injection molding are significantly higher than those determined transverse to the direction of injection molding [10,16–19]. This results in a pronounced anisotropic material behavior across the component thickness. Therefore, micromechanical approaches are often used to study the behavior of SFRT. In this context, the experimental observation and localization of damage initiation and damage propagation at micro scale remain challenging. This is why, partially contradictory results on the damage process under mechanical loading can still be found. Due to the different constituents, the following damage mechanisms occur: Matrix cracks, fiber failure and cracks in the fiber-matrix interface.

Horst and Spoormaker [20] first observe the development of voids at the fiber ends in glass fiber reinforced polyamide under tensile load. This is followed by an increase of crack growth in the fiber-matrix interface, referred to as debonding. While the quasi-static tensile load leads to a shear failure of the interface, under cyclic load the interface fails in both shear and tension. This behavior is also observed by Klimkeit et al. [21]

on short glass fiber reinforced polybutylene terephthalate and polyethylene terephthalate. In opposition to that, BELMONTE ET AL. [22] study microscope images of fracture surfaces of short glass fiber reinforced polyamide indicating the damage does not occur in the interface but in the matrix. In the in-situ investigations on short glass fiber reinforced polyamide of ROLLAND, SAINTIER, AND ROBERT [23], under quasi-static tensile loading damage initiates at fiber ends accompanied by fiber failure. Here, fiber-matrix debonding followed by a significant increase of crack growth in the matrix was only observed at relatively high loading.

2.3 Test methods

The following description introduces the principal procedure for a material characterization regarding its macro-mechanical fatigue behavior. As this experimental setup does not allow conclusions to be drawn about micro-level behavior, subsequently two experimental methods for micro-mechanical damage characterization are presented.

2.3.1 Fatigue testing

In conventional fatigue tests, specimens are loaded alternately between a constant upper and lower stress. The load is usually sinusoidal over time and the number of load cycles per second is called the loading frequency. The higher this frequency, the shorter and consequently more economical the tests. At the same time, the specimen temperature increases more with increasing load frequency, which can have an influence on the material properties. Therefore, this frequency should be limited so that a surface heating of 10 K cannot be exceeded [24].

Loading is applied either until the specimen fails or until a certain number of cycles is reached. Specimens that do not fail before this specified number of cycles is reached are referred to as runouts. The upper and lower stress, σ_{max} and σ_{min} , and consequently the mean stress, σ_m, are kept constant during a test. The ratio between the lower and upper stress is called R-ratio and constant within a test series. Each test results in one point in a lifetime or a so-called S-N diagram where the stress is plotted against the number of cycles to failure on a logarithmic scale. While it is common for metallic materials to plot the stress amplitude on a logarithmic ordinate, for fiber-reinforced plastics the ordinate often refers to the upper stress level and is plotted in linear scale. The fatigue tests can also be strain-controlled. A lifetime diagram requires testing of five valid specimens at a minimum of four stress levels. To indicate survival probability, even 24 to 30 valid specimens are necessary. The number of specimens can be reduced for preliminary studies. [24] When results are available at multiple load levels, the data are often described by a logarithmic function:

$$\sigma_{max} = a \log N + b \tag{2.2}$$

A common classification of fatigue phenomena is the distinction between low cycle, high cycle and very high cycle fatigue (LCF, HCF and VHCF). In the range of the LCF, the material is subjected to high maximum stresses. The damage behavior is similar to that under quasi-static load. The LCF refers to a number of cycles to failure of less than $1 \cdot 10^4$ or $5 \cdot 10^4$. In HCF, small elastic strain leads to failure after a high number of cycles. It is commonly defined between $1 \cdot 10^4$ or $5 \cdot 10^4$ and $1 \cdot 10^6$. The fatigue beyond $1 \cdot 10^6$ cycles is referred to as VHCF. While ductile materials such as metals have a fatigue limit in this range, fiber-reinforced plastics were not found to have a fatigue limit at all [25]. In this context, the fatigue limit refers to the stress level below which the material shows no fatigue damage and can withstand an infinite number of load cycles without failure. Since the experimental effort for this number of cycles to failure is very high, often a so-called endurance limit is defined when the material does not fail up to a number of load cycles of $1 \cdot 10^6$. Figure 2.4 shows the working principle of fatigue testing and marks the different fatigue regimes such as LCF, HCF and VHCF.

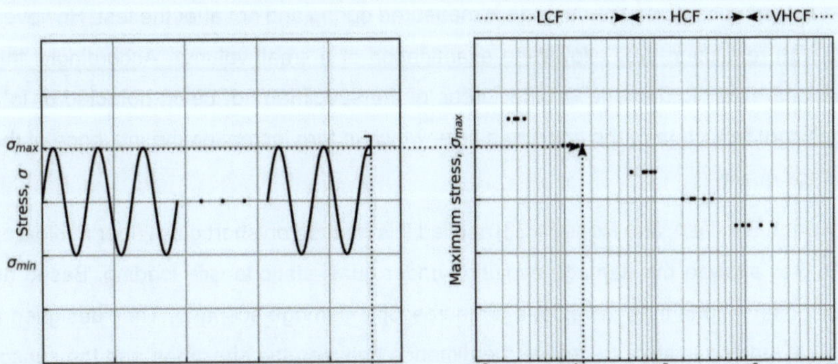

Figure 2.4: Fatigue testing for S-N diagram.

The difficulty in the cyclic material characterization of SFRT is the dependence of the mechanical properties on the loading frequency. Previous investigations suggest an increase in fatigue life with increasing loading frequency, which can be explained by a decreasing creep influence [19,26]. However, a temperature rise due to the visco-elastic behavior of the matrix ("self-heating") and due to frictional heat generated between matrix and fiber and at the fracture surface in cracks can lead to the glass transition temperature being exceeded and thus to a change in the failure mechanisms [27–29]. The influence of the frequency depends on the stress level, i.e. lower stresses are more likely to allow testing at a higher frequency. [3,30]. The use of notched specimens also allows higher test frequencies, as the maximum loaded volume is lower compared to unnotched specimens and heat is therefore transferred more quickly to less loaded areas [31].

2.3.2 Micro-computed tomography testing

Predicting the fatigue behavior of materials requires knowledge of damage initiation and propagation. Visual methods for the investigation of the fatigue damage face the challenge to achieve an adequate resolution and have the disadvantage of being mainly destructive, such as microscopic or radiographic studies. A suitable visual approach is the µCT technique that combines high spatial resolution with a non-destructive technique. For the investigation under quasi-static loading, specimens are loaded in stages and scanned after each stage. This procedure is often referred to as *in-situ* testing because the damage is measured during and not after the test. However, the high resolution only allows the examination of a small volume. Accordingly, this can lead either to damage in other areas of the specimen not being detected or to a significant reduction of the specimen size, which in turn increases the influence of the surface quality.

Rolland, Saintier, and Robert [23] applied this method on short glass fiber reinforced PA6.6 to analyze the damage evolution under quasi-static tensile loading. Based on their observations they developed a macroscopic damage scenario. They designed a custom loading system to reduce the distance between the specimen and the sensor as much as possible as this distance is decisive for the resolution. With their experimental setup, they achieved a resolution of 0.7 µm resulting in a limitation of the observation zone with a diameter and a height of 1.4 mm each. To avoid creep deformations, they reduced the scan time to about 9 min. This allowed them to scan the specimen at six strain levels before final failure.

In order to be able to investigate the damage development under fatigue loading with the µCT technique, specimens are subjected to fatigue loading outside the µCT. The loading is interrupted after a defined number of load cycles for the µCT-scan of the specimens. The main challenge here is that cracks caused by the fatigue loading close during the µCT-scan and therefore do not remain visible. This issue can be overcome by scanning the specimens while applying a predefined strain, e.g., the mean strain during the previous cyclic loading. Due to the creep tendency of the thermoplastic matrix, scanning the specimen while applying a certain strain may in turn lead to creep failure during scanning. COSMI AND BERNASCONI [32] studied the damage process under fatigue loading by scanning short glass fiber reinforced PA6.6 after applying three different numbers of cycles. Their findings suggest that a resolution of 2.0 µm is sufficient for the detection of micro-cracks within the matrix. They also address several difficulties encountered within this method, e.g., naming the limitation of material volume as critical. ARIF ET AL. [12] also analyze µCT-scans of interrupted fatigue tests on short glass fiber reinforced PA6.6 to visualize the development of damage in the fiber-matrix interface at different fatigue stages. They found out that different damage mechanisms occur in the shell and core layer. While the main damage mechanism in the shell layer is fiber-matrix debonding, they observed transverse matrix cracks in the core layer. They also point out difficulties in the application of the previously mentioned method, predominantly in achieving sufficient resolution for the detection of micro-damage.

2.3.3 Acoustic emission testing

Acoustic emission is a popular tool to gain information on the micro-structural damage mechanisms under mechanical loading. A major advantage of this method is the online measurement without further destruction of the specimen, which is why AE is referred to as a non-destructive testing method. The applicability of AE during, rather than after, loading of a component, represents the main difference from other non-destructive testing methods and appears to be its greatest advantage. For a deeper understanding of this method, this section is subdivided into three parts. At first, it describes the working principle of AE and lists the characteristics of this method, followed by a description of the application on composite materials. Subsequently the data acquisition methods are explained. Further attention is paid to the signal analysis methods since these play an important role regarding the validity of the data interpretation.

General procedure and characteristics

The AE is based on the propagation of elastic waves in a material after a sudden release of energy. Under mechanical load, an internal stress builds up. As soon as this internal stress locally exceeds the material-specific strength, crack formation occurs. Besides material heating, the energy released in this process dissipates in the form of elastic waves. Acoustic sensors that are applied on the material surface can capture these elastic waves, which propagate within the specimen, by (Figure 2.5).

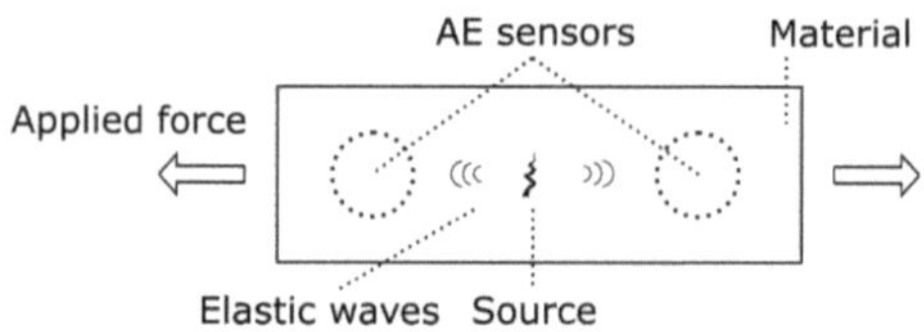

Figure 2.5: Detection of acoustic emission by a crack.

The early stages of AE took place long before sensors were used for detection. In the beginning of the 20th century, the famous case of the so-called tin cry was first documented, where the bending of a tin bar provoked a sound by internal friction due to crystal twinning that could be perceived with the unaided ear [33]. Sensors for the detection of AE were first used by JOSEPH KAISER [34], who studied the AE technique during mechanical tensile testing of various metals and reported AE for all examined metals including zinc, steel, aluminum, copper, and lead. He is therefore considered the pioneer of the AE technique. In his studies, he also observed AE during cyclic loading stating that AE activity after cyclic loading occurs only when the reloading exceeds the previous loading. This phenomenon is now known as the "Kaiser effect". In the following decades, AE became an accepted non-destructive testing method and was extended to other materials, such as concrete, asphalt and composites, and other fields of application, such as structural health monitoring [35,36]. Other than for its application for material characterization, today AE is commonly used for the online monitoring of pipes, vessels, bridges and many more [37,38].

There is a strong similarity to earthquake research, where attempts are made to detect and analyze seismic waves in order to characterize earthquakes and, if possible, to predict them. However, the geometry of the sources for the acoustic signals and thus their frequency ranges differ here by several size scales. Typically, when testing materials, waves in the frequency range between 10 kHz and 1 MHz occur [7].

In addition to the fact that AE does not require the destruction of the specimen in order to analyze the sound events, the method has numerous other advantages: Unlike the ultrasonic method, no waves are introduced into the material from the outside but the detected signal is emitted by the material itself, which is why AE testing is a so-called passive technique [39]. Another advantage of AE compared to µCT-testing (Section 2.3.2) is that it allows monitoring large structures. For this purpose, several sensors have to be arranged with distances between the sensors adapted to the material-specific attenuation. This also allows the investigation of inaccessible areas of larger components. One benefit of the sensors being highly sensitive is the possibility to capture microscopic damage events. For composite materials, this means that even the failure of a single fiber can be detected. This was first achieved by NORDSTROM in 1996, who tested matrix specimens reinforced with a single fiber and recorded the emission at the rupture of the single fiber [40].

Data acquisition

At the beginning of the measurement chain, there are the AE sensors. They are commonly made of piezoelectric elements in a protective housing that convert the surface vibration of the studied body to an electric signal based on the piezoelectric effect. They are classified in terms of their resonance behavior, which is caused by the sensor's dimensions and the piezoelectric material, into resonant and wideband sensors. While the resonant sensors show a sensitivity peak at their eigenfrequency, wideband sensors are characterized by a flat frequency response within their specific resonance range. Outside of their specific resonance range, the sensitivity of the sensors decreases significantly. [8] The advantage of resonant sensors is therefore their high sensitivity, whereas wideband sensors cover a wider frequency spectrum. The information regarding the sensor sensitivity is provided by the sensor supplier in form of calibration curves.

To ensure a complete coupling of the sensors to the solid's surface without any air pockets, the use of a coupling medium is essential. Commonly a silicon grease is used to fill gaps in the solid surface ensuring the transmission of the elastic wave to the complete surface of the sensor [41]. It is important to ensure a minimum thickness of the couplant material in order to limit the attenuation of the signals. To ensure the comparability of the AE results within a test series, the amount of the applied coupling agent should be constant in each test. The choice of coupling material can affect the measured frequency spectrum, as some materials cause a more pronounced attenuation at higher frequencies than others [42]. Besides the influence of the coupling material on the bandwidth of the signals, the influence under temperature has to be considered. While the coupling material can freeze at low temperatures and therefore cause noise, it can be subjected to degradation at high temperatures [8].

In addition to the coupling medium, the type of sensor mounting also plays an important role for the signal quality. In general, a distinction can be made between devices in which the sensors are fixed at a defined position on the specimen and devices that are attached externally or onto the testing machine. With the first mentioned, the reproducibility of the sensor position becomes challenging for several specimens. With the second type of fixation, a defined sensor position can be maintained, but friction noise can occur when the specimen moves relative to the sensor during the test. Commonly, tacky tape, clamping systems, magnetic mounts, or customized fixation systems are used for attachment.

Since the electric signals output by the sensor commonly have low amplitudes, a preamplifier is usually applied before the signal is digitalized. After processing the captured acoustic signals, the interpretation of the data can provide information on the damage state of the material [43,44]. Figure 2.6 visualizes the AE measurement chain.

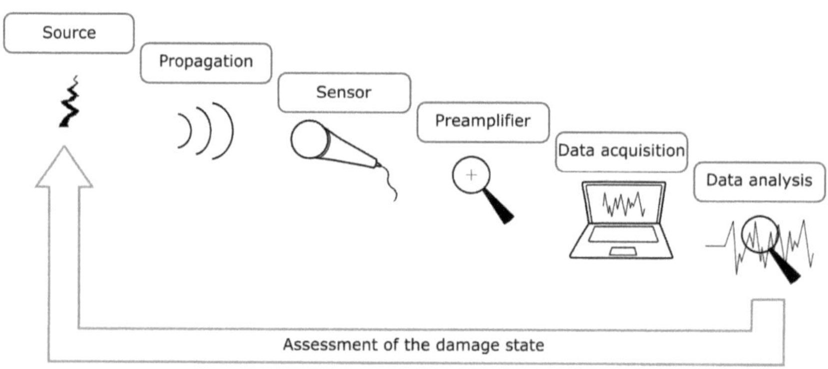

Figure 2.6: Measurement chain of acoustic emission testing.

Signal analysis

Besides capturing all relevant events, due to the high amount of data, the biggest challenge of AE lies in analyzing and interpreting the acquired information. Today most AE systems store two different kinds of datasets: the primary data and the transient data. While in the primary dataset, the AE software analyses the signals and stores parameters such as the maximum amplitude, the energy and the duration of the signal, the transient database stores the entire signals allowing a customized signal analysis. It is self-explanatory that the analysis of transient data promises to gain more information and at the same time it is associated with a larger amount of data than the analyses of only primary data. This is why, the complexity of the AE data analysis is linked to the performance of computers. It took almost 20 years from the first studies storing the entire signals in the early 1970s until the analysis of the transient signals was possible [45,46].

As it is mentioned above, the piezoelectric sensors are known for their high sensitivity. Along with the advantages of high sensitivity comes the challenge of distinguishing ambient noise from damage events inside the tested material. However, finding the best possible coupling of the sensors to the specimen is the answer to limiting this noise it requires localization of the signal's origin in order to filter out irrelevant signals that do not occur within the measurement length. Knowledge of the origin of the AE source is also important for structural health monitoring in order to repair occurring defects quickly. In material research, localized AE sources can be associated with microscopically detected damage following testing.

Determining the correct time of arrival of the signal at the sensor is crucial for accurate localization. To reduce the amount of data, the data is usually not collected continuously, but only acquired and automatically stored as soon as a certain threshold value is exceeded. Due to the similarity of the physical wave propagation, methods from earthquake research are used to determine the arrival time of acoustic signals. KURZ, GROSSE, AND REINHARDT [47] provide an overview on the most common time picking algorithms and evaluate the suitability for material research using concrete as an example. Here as well as in studies on continuous fiber-reinforced epoxy resins [48], the Akaike Information Criterion (AIC) [49] has shown to best match a manual determination of signal onset. In a first step, a Hilbert transform is applied to the signal in order to find the part, where the onset happens, as this leads to better results for the onset calculation with the AIC [50]. A simple threshold method is applied to the transformed signal to obtain an approximation of the time of arrival of the signal. The signal is reduced to several hundred samples before and after this approximated point.

Within this window, the exact time at which the signal arrives at the sensor is now estimated using the AIC according to Maeda [51]:

$$AIC(n_w) = n_w \cdot \log\Big(var\big(R_w(n_w, 1)\big)\Big) + (N_w - n_w - 1) \cdot \log\Big(var\big(R_w(1 + n_w, N_w)\big)\Big) \quad 2.3$$

where w indicates that the AIC-function is calculated for a window, n_w is each sample within the windowed signal R_w, and N_w is the last sample within the window. The first logarithmic term calculates the variance for the signal up to n_w, whereas the second logarithmic term considers the signal from n_w to N_w. The minimal AIC value within the chosen window defines the arrival time of the signal.

Besides determining the accurate arrival time of the signal, the knowledge of the speed of sound is necessary for the calculation of the source location. For a better understanding of the propagation of sound waves, some fundamental knowledge of wave propagation in solids are explained below. For further information the reader is referred to GRAFF [52], ACHENBACH [53], or SAUSE [41]. Furthermore, so-called guided waves, which are formed in bodies with finite dimensions due to boundary influences, are relevant for AE testing. These waves are often referred to as Lamb-waves, which goes back to the British mathematician and physicist HORACE LAMB, who presented an analytical formulation in 1917 [54]. Based on their displacement relative to the midplane, waves are separated into symmetric and antisymmetric modes. While symmetric modes correspond to compression or expansion waves, antisymmetric modes correlate to bending waves. Figure 2.7 visualizes the displacement in thickness direction caused by the two different modes (S_i- and A_i-mode). The index indicates the respective mode order and Lamb waves occur in at least two basic modes, S_0- and A_0-mode.

Distinguishing between the two modes is important, because the two types of waves propagate at different speeds and are frequency- and thickness-dependent. The symmetric mode usually propagates faster than the antisymmetric mode. The anisotropic mechanical properties of fiber reinforced polymers (FRP) result in anisotropic dispersion properties [55]. KELKEL [48] visualizes the dependency of the propagation velocity of the two wave modes on the direction to the fiber orientation: while the velocity of the S_0-mode is strongly dependent on the fiber orientation, the velocity of the A_0-mode shows almost no correlation with the fiber orientation. To the author's knowledge, all studies on FRP regarding the dispersion behavior of Lamb-waves have been conducted for continuous fiber reinforcement. However, since the velocity of the wave modes is directly correlated with the stiffness and therefore only indirectly dependent on the fiber orientation, it can be assumed that the findings from prior investigations can be transferred to SFRT.

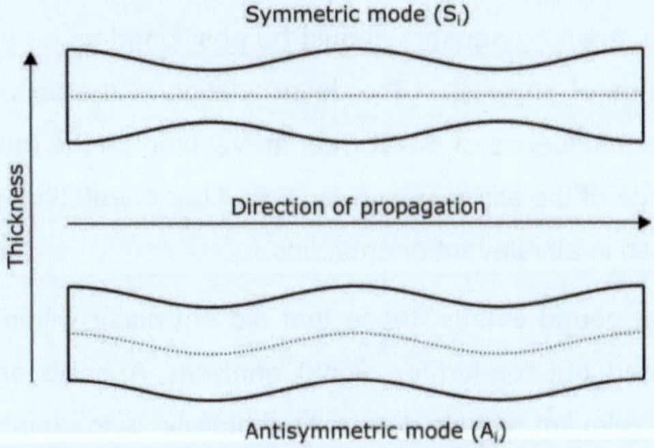

Figure 2.7: Symmetric and antisymmetric wave mode of Lamb waves with schematic displacement.

With knowledge of the exact arrival times at the different sensors and the material-specific sound velocity, the sound events can be localized. In the present work, only two acoustic sensors are used which allows a one-dimensional localization of the sound events. With the determined difference in arrival times of the two sensors, the sensor positions, and the sound velocity v, as well as the location of the acoustic event can be calculated with:

$$x_{origin} = 0.5 \cdot (t_2 - t_1) \cdot v + 0.5 \cdot (x_1 + x_2) \quad 2.4$$

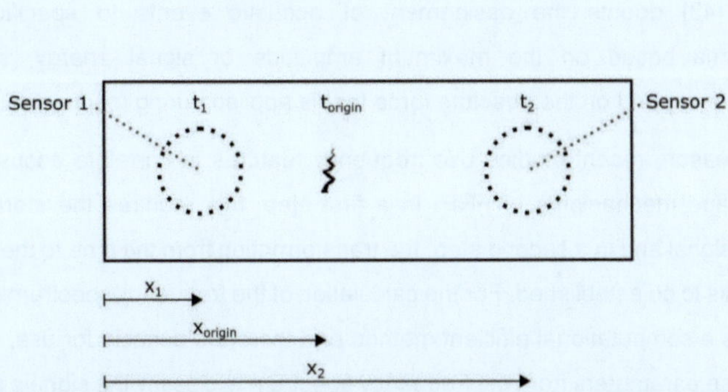

Figure 2.8: One-dimensional determination of the origin of damage event.

The speed of sound can be determined using three sensors arranged in a row, with one of the two outer sensors emitting a pulse. If the sensors are not able to emit a pulse, the Hsu-Nielsen-test can be applied, where a standardized pencil lead is broken on the specimen surface to provoke an artificial acoustic event with a broad frequency spectrum. In addition to the hardness, diameter and length of the pencil lead, the angle at which it is broken at the surface also has to be defined for the Hsu-Nielsen test [56]. In both cases, the receiving sensors should be positioned as far apart as possible to reduce the influence of an error in the determination of the arrival time. This again underlines the great influence of the correct arrival time on the quality of localization. Due to the influence of the stiffness and thus the fiber orientation, the sound velocity, must be determined in all relevant orientations.

After localizing the sound events, those that did not occur within the measurement distance are filtered out for further signal analysis. A basic analysis step, often performed for the relevant signals due to its simplicity, is to count all events. On the one hand, the acquired data is used to determine the onset of damage and on the other hand, the cumulative sum of all events is a good indicator for the damage state of the material under investigation and can be calculated without further interpreting the transient signals.

In the case of FRP, attempts are also often made to assign the recorded signals to the various damage mechanisms. Most available AE software evaluates online so-called primary parameters like maximum amplitude, signal energy, rise-time or number of threshold crossings for each signal. Particularly in the beginnings of AE testing, based on these primary parameters, the measured signals were combined into groups that were assigned to the different damage mechanisms [57–60]. However, among others RICHLER [43] doubts the assignment of acoustic events to specific damage mechanisms based on the maximum amplitude or signal energy, since both parameters depend on the absolute force that is applied during mechanical testing.

For this reason, recent studies use frequency features to correlate acoustic signals with damage mechanisms of FRP. In a first step, this requires the storage of the transient signal and in a second step, the transformation from the time to the frequency domain has to be established. For the calculation of the frequency spectrum of a signal, the FFT is a computational efficient method and therefore suitable for use. Numerous authors use parameters from the frequency spectrum to classify the signals and assign them to specific damage mechanisms of FRP [43,61–63]. For this purpose, the most frequently used frequency feature is the peak frequency, which is defined as the frequency at the maximum intensity of the signal. Since in most mechanical tests the different damage mechanisms occur partly simultaneously, pattern recognition methods are used after the relevant parameters have been determined. K-means is the most common algorithm used for the clustering of acoustic waves [64–66].

However, the author is only aware of studies that link AE sources to damage mechanisms of continuous fiber-reinforced plastics. KELKEL [48] provides an extensive overview on studies that link the peak frequency to different damage mechanisms of continuous fiber-reinforced plastics. His comparison shows the trend that matrix cracks can be assigned to signals with low peak frequency and fiber breaks to signals with high peak frequency. Nevertheless, the absolute values vary and therefore seem to depend on the material and the experimental set-up. This finding also agrees with the results NORDSTROM [40] presents by characterizing the acoustic waves generated in single-fiber fragmentation test: the larger the distance between the sensor and the source, the greater the proportion of low frequencies. The sensor used also influences the frequency spectrum of a signal, as each sensor has different sensitivities in certain frequency ranges.

Inaccuracies

In addition to the inaccuracies mentioned above, such as the sensor-source distance and the sensitivity of the sensors used, there are other factors that must be taken into account when interpreting AE data. The definition of the threshold value, after which an acoustic event is recorded, is of great importance here. This threshold value is set, so that background noise is already filtered out during the measurement. Of course, this also involves the risk that relevant acoustic events with low amplitude are not recorded. Furthermore, too high material damping and too large sensor distances can result in damage events not being detected. As previously mentioned, non-uniform or insufficient coupling of the sensors to the specimen, can lead to further measurement inaccuracies.

Usually, preamplifiers are used to amplify the signal. This may result in an overdriven signal for which an analysis is not possible or only possible to a limited extent, since information about the exact signal path is missing as soon as the signal exceeds the maximum measurable amplitude.

In addition to the definition of a threshold value, a minimum and a maximum signal length are defined, which creates another possible source of error. If the signal length is selected too short, signal information is lost, if a too long signal duration is chosen, overlapping of successive signals may occur.

Another inaccuracy in the measurement of acoustic signals results from the selected sampling rate. Therefore, a balance must be found between a realistic amount of data and sufficient accuracy in signal acquisition.

Application on SFRT

In the discussion above, on signal analysis, some applications of AE on continuous fiber reinforced polymers were listed. To the authors knowledge, AE on SFRT was first studied by SATO ET AL. [60] in the 1980s. In their tests, crack initiation and propagation in the matrix and the fiber-matrix interface were distinguished based on the amplitude of the captured acoustic signals. Low-amplitude signals were assigned to both crack initiation and growth of micro-cracks in the interface. High-amplitude acoustic events were associated with the occurrence of matrix cracks. BARRE AND BENZEGGAGH [57] also used the maximum amplitude to link signals to the different damage mechanisms of short glass fiber reinforced polypropylene. They validated their findings by scanning electron microscopy observations and hereby linked low amplitude events to cracking in the matrix and in the interface. Medium amplitude events were associated with fiber pull-out and high amplitude events to fiber failure. In *in-situ* tests (see Section 2.3.2) combined with AE testing on short glass fiber reinforced polypropylene and polybutylene, the damage mechanisms during tensile loading were studied by evaluating the acoustic energy [67]. The acoustic energy is a parameter that is related to the maximum amplitude, but dependent on the window selection of the signal, which is why the author doubts the significance of this parameter.

A few publications that are more recent study the features extracted from the frequency domain during AE testing of SFRT. BENZ, WILLEMS, AND BONTEN [68] carry out tensile tests with acoustic emission on short glass fiber reinforced polycarbonate. They use the peak frequency to assign the registered acoustic signals to the three damage mechanisms matrix, interface and fiber fracture. Furthermore, they correlate the onset of AE with the first damage detection and thus define a so-called critical strain below which no damage is supposedly initiated. A similar approach was used by BAUER ET AL. [69] to derive a correlation between the acoustic signals measured in the quasi-static tensile test and the fatigue behavior of short fiber reinforced PA.

2.4 Digital twin simulation

Simulations of the material behavior prove challenging due to the complex microstructure of SFRT resulting from the injection molding manufacturing process. In addition to statistical distributions such as the fiber length and fiber orientation distribution, the nonlinear behavior of the thermoplastic matrix must be taken into account. Depending on the problem being investigated, different modeling approaches can therefore be chosen, which differ in their depth of detail. A distinction is typically made between three levels: macro, meso and micro level simulation [70–72]. Commonly, simulations on all levels use the finite element method (FEM).

2.4.1 Macro-mechanical analysis

Macro-mechanical analysis is used to study the mechanical behavior on component level. It aims to identify regions of stress concentrations and study the deformation behavior. At this level, the properties of fibers and matrix are blended to a homogeneous material either with isotropic or anisotropic properties. If experimental characterization of the required homogeneous material properties is not possible, they can be estimated using micromechanical approaches. The basic work on the calculation of the reinforcing effect of an ellipsoidal inclusion on the pure matrix properties was published by ESHELBY [73] as early as 1957. His work was followed by numerous further developments, such as by MORI AND TANAKA [74], who extended Eshelby's findings to include the interaction of multiple fibers. Because of its simplicity, the model of HALPIN AND TSAI [75] is still commonly used today to calculate the elastic constants of SFRT. TUCKER AND LIANG [76] give a comprehensive review of different models for predicting the stiffness of SFRT and conclude that the predictions of the Mori-Tanaka model are among the best.

2.4.2 Meso-mechanical analysis

On the meso-scale, a more detailed study can be performed than with a macro-mechanical analysis. Models on this scale are often used for the analysis of laminates with different layers [77,78]. These models map the different laminate layers and the interfaces between the layers to subsequently predict the macro-mechanical properties, as well as failure and failure progression. Meso-scale models of SFRT consider the fiber orientation distribution by mapping the fiber orientation tensors to the finite element mesh [79,80]. For each element, the elastic constants are calculated based on the respective fiber orientation. This approach allows to take into account the complex anisotropy of SFRT, but does not allow to draw conclusions about the damage process in terms of the different damage mechanisms such as matrix cracks or fiber failure.

2.4.3 Micro-mechanical analysis

A micro-mechanical simulation can provide prediction of some damage mechanisms. Since simulations on micro-level distinguish between the different material components, the computational power only allows the simulation of a limited volume. When using the FEM, the mechanical behavior of such heterogeneous materials is described by means of so-called representative volume elements (RVEs). HILL [81] published the first definition of an RVE in 1963, using it to denote areas of reinforced solids that are statistically representative of the entire composite independent of its defined boundary conditions. Another frequently cited definition comes from DRUGAN AND WILLIS [82], who limit the size of an RVE to the smallest volume that allows calculation of an effective average of the composite's properties.

Although RVEs allow the consideration of statistical distributions such as the fiber length distribution or fiber thickness distribution, both the calculation of complex RVEs and the determination of the statistical input variables is very time-consuming. As a result, in most studies on RVE's, numerous simplifications of the microstructure are performed. A comprehensive overview of the influence of different simplifications on the prediction quality of an RVE is given by BREUER AND STOMMEL [83]. They highlight the importance of the fiber packing for the predictive quality of the composite properties. In addition, they find the fiber length distribution and the fiber orientation to play an important role. Furthermore, the fiber aspect ratio within the RVE has an important influence on the strain hardening behavior of the predicted stress-strain curve [84].

Microstructural simulations aim to determine the local stresses and strains occurring in the matrix and the fibers in order to study the interaction of the different damage mechanisms during the damage process. Recent studies also incorporate the fiber-matrix interface into the simulation. In addition to an enormous computing capacity, the challenge here lies particularly in the experimental characterization of the interface properties [85,86].

In order to investigate the damage mechanisms of SFRT under tensile loading, FLIEGENER, KENNERKNECHT AND KABEL [87] limit the dimensions of their specimens so that their entire microstructure can be accurately captured by µCT scans. They perform comparative simulations using FE simulation and the commercial software GeoDict® [9]. This software runs the calculation on a voxel grid and uses computationally efficient FFTs. The main advantage of these models compared to FE models is that no mesh is required and that the simulation can be performed directly on the µCT images. This is the reason why these models are called digital twins. The numerical implementation of the computation of the mechanical response of heterogeneous materials based on microstructural images was introduced by MOULINEC AND SUQUET in 1994 [88]. They develop an iterative algorithm using the Fourier transform to solve the so-called periodic Lippman-Schwinger equation [89]. Further information on the FFT-based technique for the computation of the mechanical response of a composite is provided in [90–92].

3 Problem statement and objective

The previous chapter listed the various influence factors on the mechanical properties of SFRT. Therefore, a material characterization that takes into account all influencing factors is associated with a high testing effort. This testing effort becomes even higher as soon as it is a matter of characterizing the fatigue properties, since fatigue tests are very time-consuming in themselves. Despite the high cost of fatigue tests, existing research mostly examines the influence of various factors on fatigue strength [3,10,11,17,19], and the information is often used only to create another point on the S-N diagram instead of drawing conclusions about the damage process and the underlying damage mechanisms. Moreover, in the case of SFRT, the few studies on the damage process under fatigue loading show partially contradictory results [20–23].

Therefore, this work aims to investigate and develop methods that allow a time-efficient investigation of the fatigue behavior of SFRT. It is focusing on exploiting the opportunities offered by data analysis methods. For this purpose, not only is the final failure documented in the form of an S-N curve, but also other available descriptors such as acoustic emission signals, mechanical hysteresis and stiffness degradation are also used to form a complementing dataset, which is then analyzed with respect to the S-N curve.

Machine data under cyclic loading is readily available and therefore worth evaluating. The advantage of using this data is that it requires no further test equipment. Machine data is used to assess stiffness degradation and hysteresis behavior during cyclic loading. The choice ultimately landed on these two parameters as they are directly linked to the damage behavior in the material. These parameters are then analyzed to gain further insight into the fatigue behavior. It is also investigated whether a loading with only a limited number of cycles is sufficient to obtain the same information and allows to parameterize a model that allows prediction of the material behavior under progressive cyclic loading. This could reduce the testing time and thus the costs for the material characterization.

In a second step, the possibilities offered by AE testing are investigated. The approach for AE testing in this work is based on the assumption that damage modes sequence under quasi-static load is similar to that introduced by fatigue loading: in the case of quasi-static loading, the damage increases with increasing load or dissipative energy.

With cyclic loading, the deformation energy to which the material is subjected also increases over the life of the material, leading to an increase in damage. Thus, under both types of loading, there is a relationship between increasing deformation energy and increasing material damage.

Due to the long run times and the enormous amount of data generated, it is not practical to apply AE directly in fatigue tests. Therefore, to again reduce testing time and effort, it was the author's intention to establish a link between static and fatigue behavior.

Within the presented work, quasi-static tensile tests with acoustic emission are conducted first on unloaded and second on cyclically preloaded specimens up to a defined number of load cycles, with the objective of comparing the acoustic behavior. The idea is that damage, which has already occurred during preloading, cannot be monitored during quasi-static residual strength testing, because the AE signals are related to the release of strain energy due to new or evolving damage. Consequently, if the acoustic response of the non-preloaded and preloaded specimens matches, the cyclic loading did not cause any damage, which is an indication of a limit for damage initiation.

Finally, the statistical analysis of the available data is tested together with the underlying assumptions using a micromechanical model. Assuming that observations on damage behavior under quasi-static load allow conclusions on behavior under fatigue loading, micromechanical simulation offers the possibility to gain insight into the different damage mechanisms. Therefore, as a third aspect of this work, a digital twin simulation is built to verify the above-explained experimental investigations by micro-level observations.

To ensure comparability of the results of the various investigations explained above, a uniform specimen geometry is to be used for all the tests. This results in special requirements on the specimen geometry, which are explained in more detail in the following chapter. Subsequently, the investigations within the fatigue tests are presented followed by the study of the AE behavior after fatigue preloading compared to the AE behavior of neat material. Finally, the simulation of the behavior at the microstructural level is discussed. The structuring of the content follows the sequence shown in Figure 3.1.

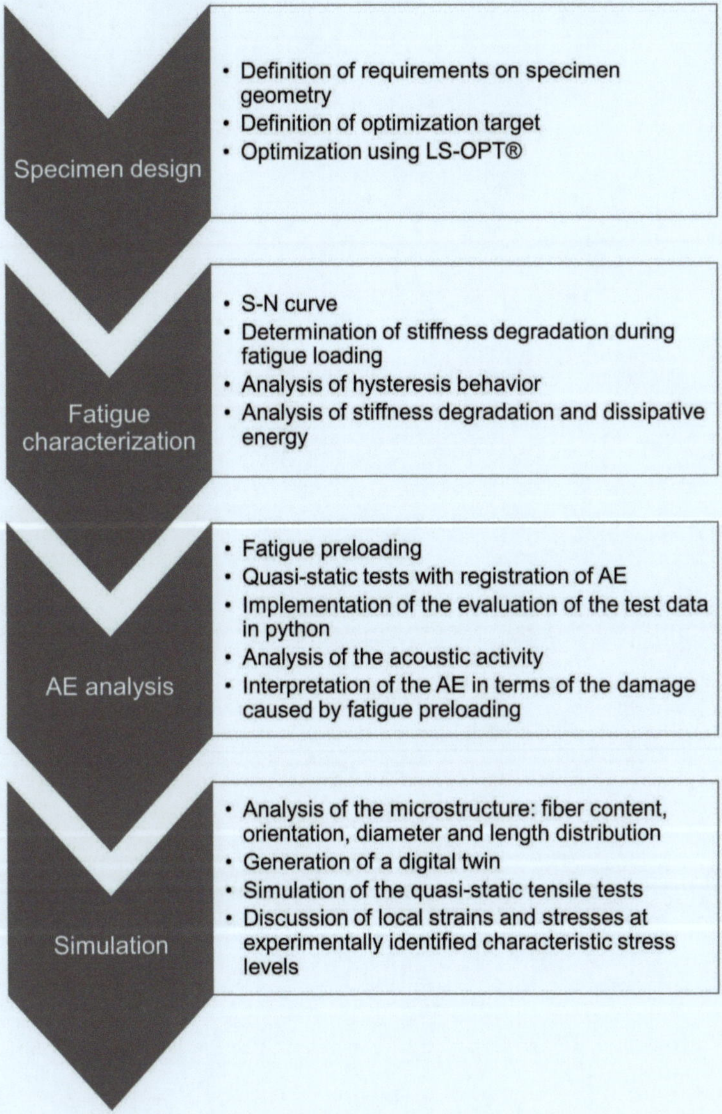

Figure 3.1: Procedure and structure of the present work.

4 Design of specimen geometry

4.1 Material

The material under investigation is a short fiber reinforced polyamide 6.6 (Ultramid® A27 E, BASF SE, Ludwigshafen, Germany) containing 40 wt.-% glass fibers. The granules for the production of the test specimens were provided by Lehmann&Voss&Co KG, which used an E-glass fiber of Nippon Electric Glass Co., Ltd (CAS: 65997-17-3) for reinforcement. Plates (80 x 80 mm^2) with a thickness of 3.0 mm were produced by injection molding, from which the test specimens were subsequently taken by milling. See Appendix A.1 for the parameters of the injection molding process.

4.2 Optimization

The specimen geometry was developed with the following requirements in mind:

(1) Use of the existing mold to save time and cost leading to restrictions on the specimen size;

(2) Evaluation of the microstructure via X-ray microscopy leading to restrictions on the dimensions of the gauge length;

(3) Application of the acoustic sensors as close as possible to the measurement length leading to requirements on the available surface;

(4) Minimization of stress concentrations.

The optimization of the specimen geometry was performed with the software LS-OPT® from DYNAmore. During the optimization process, the stresses were determined using the finite element analysis software Abaqus® with a transverse isotropic material law. The material properties were taken from the work of BAUER [93], since here, in addition to the elastic constants and the tensile strengths at 0° and 90° to the injection molding direction, also the shear parameters are given and thus all the necessary parameters are available. The Young's modulus, the Poisson ratio, and the strength under tensile loading were determined on PA6.6 with 30% glass fiber reinforcement and for specimens with a thickness of 2 mm as presented in Table 4.1. Although the fiber content and the specimen thickness are different in the present work, the comparability with the material properties is nevertheless expected under the assumption that the influence of these two factors on the degree of anisotropy balances out. On the one hand, higher fiber content results in a higher degree of anisotropy, on the other hand, a higher specimen thickness leads to a decreased degree of fiber orientation.

In the following, it is specified what these four requirements mean and how they were implemented: the existing mold has the dimensions of 90 mm x 80 mm. The specimen length was therefore limited to 75 mm, so that machining of the edges with a milling cutter remains possible also for specimens transverse to the direction of injection molding.

Table 4.1 Mean value of tensile properties of PA6.6GF30 [93]

Θ to direction of injection molding in °	E_t, respectively. G in MPa	σ_t respectively τ in MPa	ν
0	6068	104	0.47
45	3578	44	-
90	3198	65	0.20

The resolution was limited to 3.0 µm to resolve a single glass fiber with multiple voxels in the X-ray scan. With this high resolution, the available X-ray allows to scan a volume with the edge length limited to 8 mm, so the measurement length was defined to 8 mm. The width of the specimen, on the other hand, was set at 3 mm in order to obtain a square cross section and limit the computational effort in the microstructural simulation. The high resolution also requires a limitation of the clamping width, so that the sample can rotate close enough to the detector in the X-ray. The clamping width was then set to a maximum of 22 mm.

As a third requirement, the sensors must be applied with the complete piezo-element on the specimen surface, which means that signal acquisition with a sensor diameter of 8 mm cannot take place directly in the test cross-section. Nevertheless, the sensors are to be placed as close as possible to the measuring length to limit attenuation of relevant signals. Programmatically, this was implemented by defining a constraint that limits the ratio of the distance between the sensor and the center of the specimen and the distance between the sensor and the clamping in the testing machine to 1.

With a defined specimen width of 3 mm, a sensor diameter of 8 mm, and a clamping width of 22 mm this leads to a significant widening of the cross section. Since the material behaves very sensitively to stress concentrations, but failure is to be provoked by internal or microstructural stress concentrations rather than by external stress concentrations, the smoothing of this transition from narrow to wide cross-section is chosen as the objective of the optimization. This results in a uniform stress state in the measurement length avoiding geometric stress peaks. Implementing this optimization objective means minimizing the ratio between the maximum stress and the average stress within the measurement length. Therefore, the parameters of a quadratic Bézier curve were optimized for a smooth transition from the narrow to the wide cross section (see Figure 4.1).

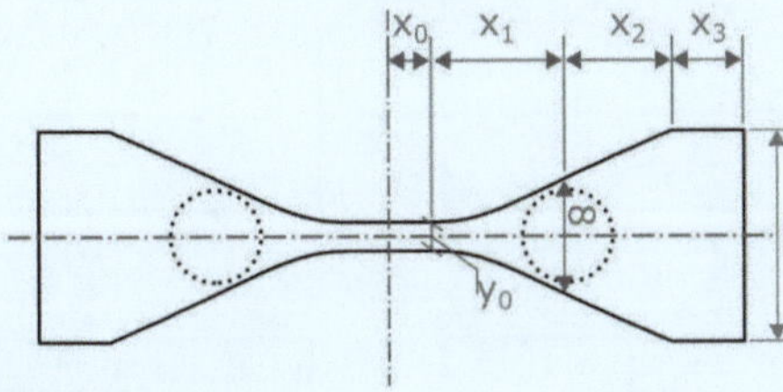

Figure 4.1: Optimization parameters.

Widening of the cross-section is associated with curvature of the contour. High shear stresses occur at these points. Another requirement for the optimization of the specimen geometry was therefore to avoid failure due to excessive shear stresses or an excessively high combination of shear and transverse tensile stresses. This was estimated with the calculation of the so-called material effort. In order to calculate the material effort, the Tsai-Hill failure criterion was used by means of a user-subroutine. It has been successfully applied as failure criterion for SFRT [3,94] and is defined by:

$$F_{Tsai-Hill} = \left[\left(\frac{\sigma_1}{\sigma_{1,u}}\right)^2 - \frac{\sigma_1\sigma_2}{\sigma_{1,u}^2} + \left(\frac{\sigma_2}{\sigma_{2,u}}\right)^2 + \left(\frac{\tau_{12}}{\tau_{12,u}}\right)^2\right] = 1 \quad 4.1$$

Where:

$\sigma_1/\sigma_{1,u}$ is the stress / strength in the injection molding direction (0°)

$\sigma_2/\sigma_{2,u}$ is the stress / strength transverse to the injection molding direction (90°)

$\tau_{12}/\tau_{12,u}$ is the in-plane shear stress / strength between the 0° and the 90° direction.

While a material effort $F_{Tsai-Hill}$ of 1 is obtained within the gauge length, it is limited to a value of less than 1 in the rest of the specimen. Thus, the failure of the specimen occurs within the gauge length. Figure 4.2 shows the optimized specimen geometry and visualizes the different requirements for the optimization process.

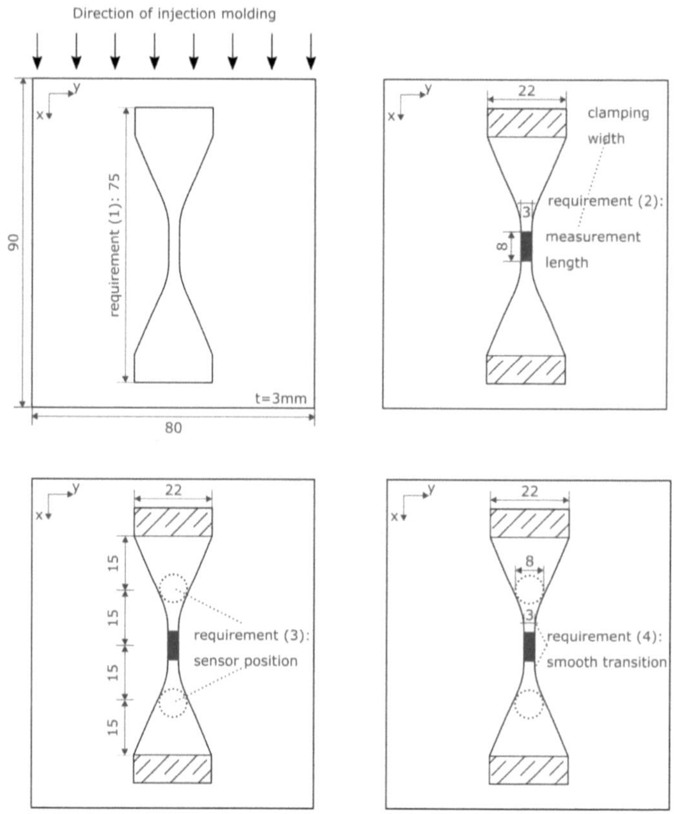

Figure 4.2: Optimized specimen geometry with requirements for the optimization process.

To shorten the calculation time, the specimen symmetry was used in the finite element analysis and the stresses were calculated for only a quarter of the specimen. Shell elements with four nodes and reduced integration elements with a mesh size of 0.2 mm were used (Abaqus® element type S4R). A study on the mesh sensitivity was not conducted. Instead, the most promising four different specimen designs were identified after the optimization process and the final specimen design was chosen based on the results of preliminary tests with these different specimen geometries. The further testing was conducted with the specimen geometry where most specimens failed within the measurement length. Table 4.2 lists the geometric parameters and their constraints with reference to Figure 4.1. The stress distributions of the optimized specimen geometry under pure tensile loading can be found in A.2.

Table 4.2 Geometric parameters of optimized specimen geometry

Parameter	Value	Constraints
x_0	4.0 mm	3.0 – 4.0 mm
y_0	1.5 mm	fixed
x_1	12.3 mm	9.4 – 15.0 mm
x_2	13.5 mm	10.0 – 15.6 mm
x_3	7.7 mm	6.4 – 10.0 mm

5 Fatigue testing

The fatigue testing was carried out on an electrodynamic testing machine with a load cell of 15 kN calibrated to 0.5% full scale (ElectroForce 3550, TA Instruments, USA). The tensile-tensile, sinusoidal load with a load ratio of $R = \sigma_{min}/\sigma_{max} = 0.1$ was applied with a constant testing frequency. A temperature chamber was used to assure a constant temperature of 23 °C. Under cyclic loading, the visco-elastic material behavior and internal friction lead to a temperature rise. Since the material behavior depends on temperature variations, the international standard ISO 13003 [24] restricts the surface heating to 10 K.

In a preliminary study, the maximum permissible test frequency was determined to be 10 Hz, so that the maximum permissible specimen heating was not exceeded. The specimen surface was observed with an infrared camera (PI400, Optris, Germany) to monitor the temperature increase. Figure 5.1 shows the heating of the specimen surface over the number of cycles plotted in logarithmic scale for different maximum stress levels (σ_{max}). All three curves show qualitatively the same course. While the largest temperature increase occurs at the beginning, the sample almost does not heat up during the rest of the test until a steep temperature increase is observed shortly before failure. Both the absolute temperature increase and the number of cycles until a quasi-constant temperature is established depend on the maximum stress. The higher the maximum stress, the stronger and faster the temperature rises. However, the maximum temperature increase is below 6 K even at the highest maximum stress level. This is only true for tests in the high and the very high cycle regimes. For maximum stresses that lead to failure in low cycle fatigue, the surface temperature increase exceeds the limit of 10 K for the chosen testing frequency of 10 Hz. However, since the focus of this work is on the range of high numbers of load cycles, and these can only be achieved with justifiable effort at a high test frequency, this temperature increase was accepted.

After the specimens had been stored in a room climate for approximately 18 months, one part was conditioned in a climate chamber (ClimeEvent, Weiss) at a temperature of 70 °C and a relative humidity of 40 % (RH40) which corresponds to the moisture content of the lab air during the experiments. This showed that mass constancy had already been achieved. The moisture content during the tests was measured with a hygrometer (Testo 174-2010, Testo SE & Co. KgaA, Germany) to ensure stable and similar conditions for the entire test series. Test results under deviating environmental conditions were not evaluated as temperature and humidity have an important impact on the mechanical properties.

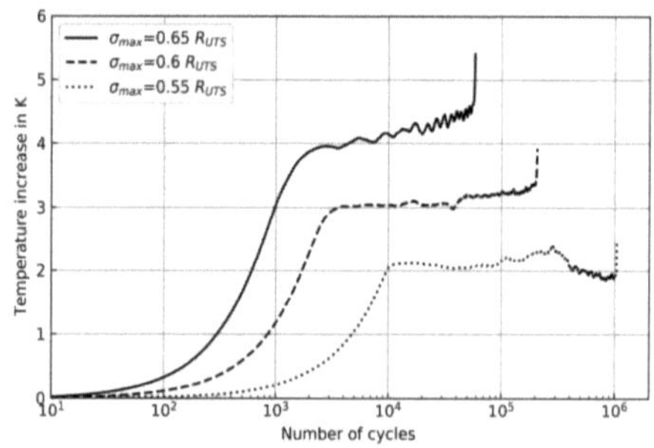

Figure 5.1: Specimen surface heating during cyclic test with different σ_{max} (R=0.1).

5.1 S-N curve

For the determination of the fatigue life tension-tension (R = 0.1) tests were carried out under load-control. The specimens were tested until failure or until reaching 10^7 load cycles. Those that did not fail before reaching 10^7 load cycles are marked as runouts.

Table 5.1 Test matrix of fatigue loading for determination of S-N data

0° to direction of injection molding			90° to direction of injection molding		
Maximum stress in MPa	Relative maximum stress (σ_{max}/R_{UTS})	N° of specimens	Maximum stress in MPa	Relative maximum stress (σ_{max}/R_{UTS})	N° of specimens
93.1	0.62	3	68.6	0.60	3
86.5	0.57	5	65.4	0.57	3
79.8	0.53	3	60.4	0.53	3
73.2	0.48	3	54.9	0.48	3
66.5	0.44	3			

Two test series were conducted, one with specimens cut in the direction of injection molding (0°) and a second one with specimens cut transverse to the direction of injection molding (90°). Figure 4.2 shows the chosen specimen geometry. Since fatigue tests show important scatter, multiple specimens were tested at different stress levels ranging between 62% and 40% of the ultimate tensile strength. The tests in the second series of experiments with 90°-specimens were for verification purposes and therefore less extensive. Table 5.1 summarizes the test matrix for the determination of the S-N data.

The S-N curve for the 50%-survival probability ($P_S = 50\%$) is determined with a *python*-code. The non-linear least squares method is applied to fit the data to Eq. 2.2. Table 5.2 shows the calculated parameters a and b for both the 0°- and the 90°-specimens. The calculation does not take into account quasi-static tensile strength or runouts.

Table 5.2 Parameters a and b of cyclic tests for 0°- and 90°-specimens (R = 0.1, f = 10 Hz)

Angle to direction of injection molding	Absolute S-N curve		Relative S-N curve	
	a	b	a	b
0°	-9.12	128.80	-0.060	0.849
90°	-7.2	95.9	-0.063	0.840

Figure 5.2 shows the test results and the calculated S-N curves. For the 0°-specimens, the approximation of the data underestimates the quasi-static tensile strength by approximately 25 MPa. For the 90°-specimens, the quasi-static strength is underestimated by approximately 15 MPa. It can be observed that the S-N curve for the 90°-specimens runs slightly steeper than the one for the 0°-specimens. The fatigue strength transverse to the injection molding direction is on average approx. 25% below the fatigue strength in the injection molding direction.

Figure 5.3 shows the number of cycles to failure at different stress levels normalized to the quasi-static tensile strength R_{UTS} (see Section 6.1.2 for the quasi-static tensile testing conditions). It should be noted that the testing velocity for the cyclic tests differs from the testing velocity for the quasi-static tests. Due to the constant frequency in the cyclic tests, the test speed ranges between 35 mm/min and 160 mm/min. Although previous studies on SFRT show an influence of the testing velocity on the stiffness and strength, this influence factor was not considered in the present work due to the wide range of the testing velocity in the cyclic tests [95,96]. With normalizing the cyclic maximum stress to the respective quasi-static tensile strength, both curves show a similar slope.

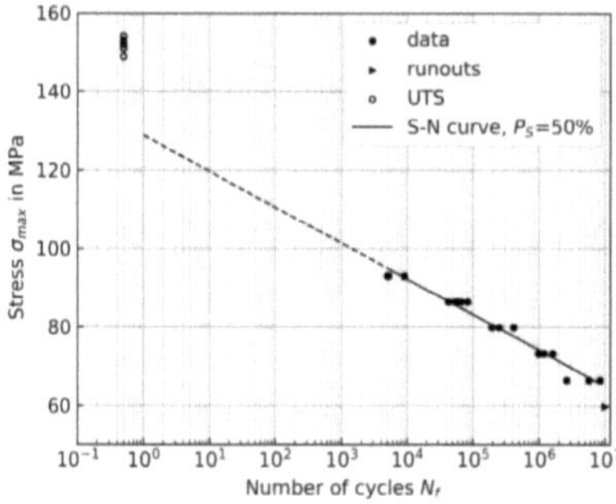

(a)

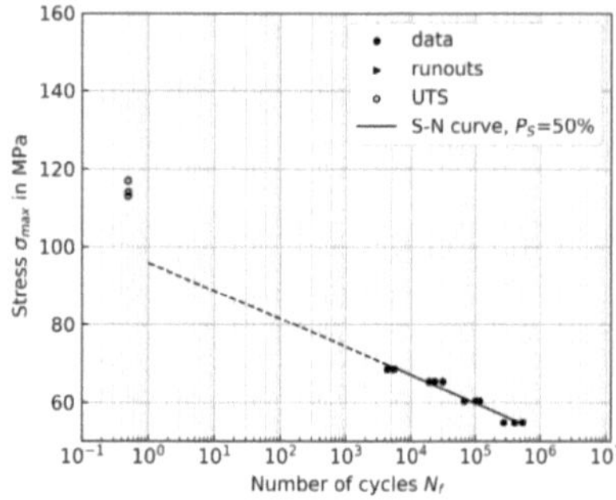

(b)

Figure 5.2: Results of the S-N tests: (a) 0°-specimens; (b) 90°-specimens.

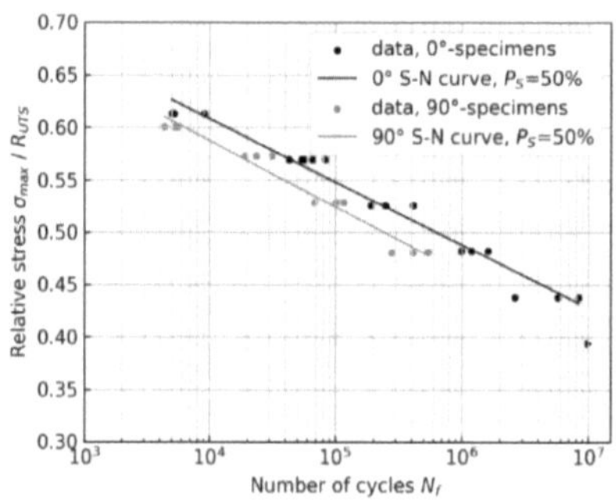

Figure 5.3: S-N data and curves normalized to the quasi-static tensile strength for 0°- and 90°-specimens.

5.2 Stiffness degradation

As appearing microcracks and fiber failure reduce the stiffness, the observation of a stiffness degradation is useful to evaluate the overall damage development during fatigue loading. At the same time, the stiffness of the material can be evaluated without further test equipment by analyzing the machine data, such as the applied load and the machine's crosshead displacement. Most available control software offers the possibility to store the data in different formats. For the evaluation of the stiffness degradation the so-called peak-valley data is used, which stores the applied force and displacement at each peak respectively to each valley (see Figure 5.4).

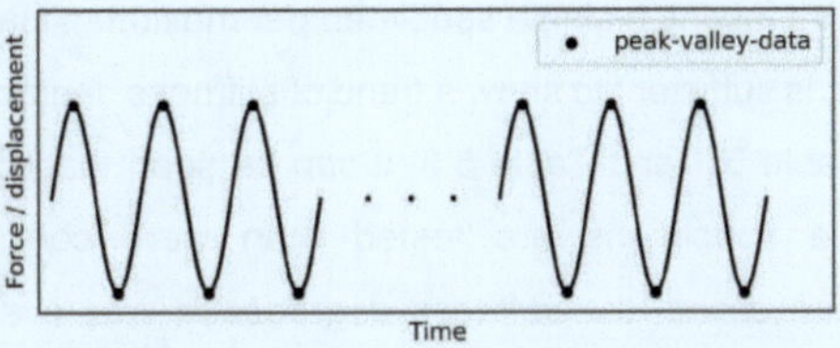

Figure 5.4: Data acquisition for analysis of stiffness degradation.

The dynamic stiffness of each cycle is calculated by dividing the force difference by the displacement difference:

$$E_{dyn} = \frac{F_{max} - F_{min}}{s_{max} - s_{min}} \qquad 5.1$$

Relative stiffness was used for further considerations to facilitate comparability of the results. For this purpose, the stiffness was referenced to the stiffness at a load cycle number of 100, as the machine needs a certain time to reach a steady state. Table 5.3 lists the test matrix for the investigations on the stiffness degradation.

Table 5.3 Test matrix of fatigue loading for determination of stiffness degradation

0° to direction of injection molding			90° to direction of injection molding		
Maximum stress in MPa	Relative maximum stress	N° of specimens	Maximum stress in MPa	Relative maximum stress	N° of specimens
93.1	0.62	3	70.5	0.62	1
86.4	0.57	3	68.6	0.60	1
79.8	0.53	3	66.0	0.57	1
75.5	0.50	1	62.9	0.55	1
73.2	0.48	3	60.4	0.53	1
66.5	0.44	3	57.2	0.50	1
59.8	0.40	3	51.4	0.45	1
53.2	0.35	1	45.3	0.40	1
46.6	0.31	1	40.0	0.35	1
39.9	0.26	1	34.3	0.30	1
33.2	0.22	1	28.6	0.25	1
30.2	0.20	1	25.2	0.22	1
26.6	0.18	1	22.9	0.20	1
20.0	0.13	3	20.1	0.18	1
			15.1	0.13	1

It should be noted that testing a single specimen per maximum load level, although not statistically relevant, is sufficient to show a trend of stiffness degradation. By comparing the test matrix in Table 5.1 and Table 5.3, it can be seen that at some of the higher stress levels, more specimens are tested than were considered for stiffness degradation. In these cases, the stiffness degradation was evaluated for all tested specimens and only the specimen closest to the mean stiffness degradation was considered in the further evaluation, thus avoiding overweighting of individual stress levels.

In the following, the procedure of the investigations for the specimens cut with 0° to the injection molding direction (0°-specimens) are shown. Usually, the stiffness of the material is calculated using strain measurement, which requires a more complex test setup though. However, taking the crosshead displacement of the machine to calculate stiffness results in significant noise of the signal. Therefore, the stiffness is smoothed before further data analysis using a *python* code. Since any fatigue parameters are usually represented by a logarithmic scale, the first step was to downscale the number of load cycles logarithmically. This leads to a complete consideration of the first part of the data and a noise reduction only for the high cycle data. In a second step a rolling window calculation is applied, which is part of the *python pandas* library. The size of the moving window was set to 100 data points. Figure 5.5 shows exemplarily the relative and the smoothed relative stiffness of a 0°-specimen with a maximum stress of (a) 86.4 MPa and (b) 73.2 MPa.

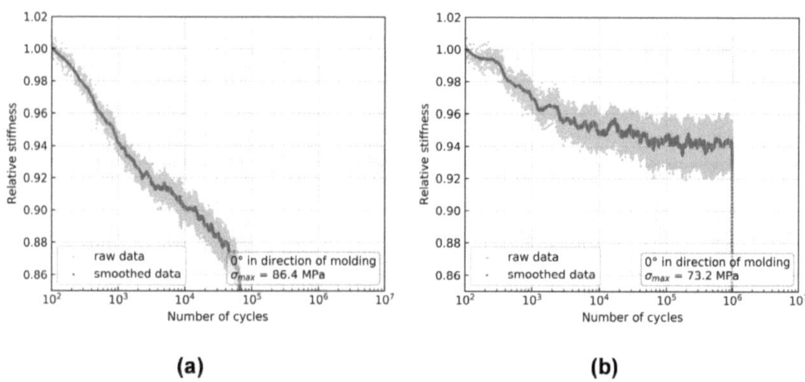

Figure 5.5: Relative stiffness data and smoothed relative stiffness of 0°-specimen: (a) σ_{max} = 86.4 MPa, resp. 0.57 R_{UTS}; (b) σ_{max} = 73.2 MPa, resp. 0.48 R_{UTS}.

Figure 5.5 shows that the stiffness degradation can be divided into three sections. While there is a significant stiffness degradation at the beginning of the test, the stiffness degradation flattens during most of the lifetime. A significant drop in stiffness precedes the final failure. This curve progression is similar to the specimen heating shown in Figure 5.1. In contrast to the temperature heating curves, the change in the slope of the stiffness degradation was found to occur between 10^3 and 10^4 load cycles regardless of the maximum stress. This was observed in all specimens loaded with a maximum stress of at least 26.6 MPa. Below this maximum stress, no stiffness degradation was observed using the machine data.

It was found that the relative stiffness degradation until the final drop can be expressed by two logarithmic function, where N is the number of cycles:

$$Relative\ stiffness = a \cdot \log_{10} N + b \tag{5.1}$$

Figure 5.6 shows the logarithmic fitted curves for the same specimen as in Figure 5.5 with a maximum stress of 73.2 MPa. The curve parameters were determined using the *SciPy* library of *python*.

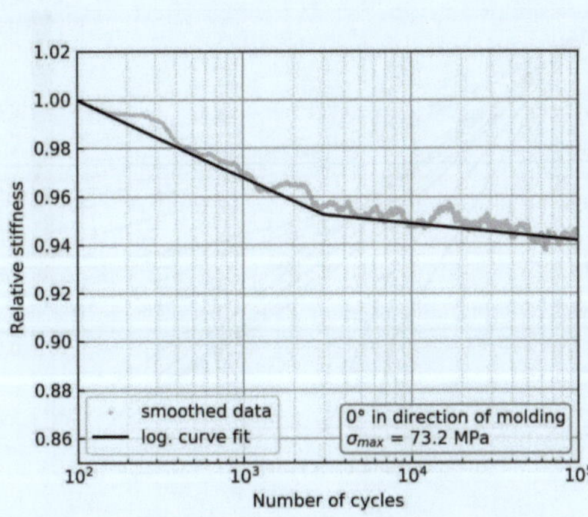

Figure 5.6: Smoothed relative stiffness and fitted logarithmic curves of 0°-specimen, σ_{max} = 73.2 MPa, resp. 0.48 R_{UTS}.

In a next step, the aim is to investigate how many load cycles are required as a minimum to determine the stiffness degradation curves with the same parameters in order to shorten the test times. For the specimen shown in Figure 5.6 and all other specimens, it was found that there was no further change in the curve fit parameters when only the data up to a load cycle number of 10^5 was included. Since this significantly reduces the amount of data on the one hand, and the test duration on the other hand, in the further processing the stiffness degradation data were only evaluated between 10^2 and 10^5 load cycles.

Figure 5.7 summarizes the fitted stiffness degradation curves for all 0°-specimens. The grey scale indicates the maximum stress level with decreasing stress from black to white. It can be observed that the change of slope of the stiffness degradation occurs at about 10^3 load cycles or in other words, that the stiffness degradation can be divided into two sections: below and above 10^3 load cycles. In the first section, the stiffness decreases faster than in the second section. This behavior is observed for all the specimens subjected to a maximum stress between 20% and 60% of the ultimate tensile strength (R_{UTS}). For the specimens loaded with 62% of the R_{UTS}, this turns around and the stiffness decreases faster in the second part of the loading history. Interestingly, this is the only load level in the range of LCF, which is commonly defined for a number of load cycles up to 10^4 cycles.

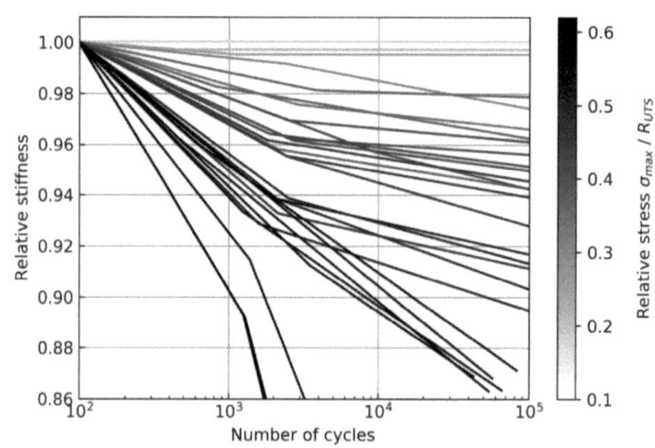

Figure 5.7: Fitted relative stiffness curves of 0°-specimens.

It can be observed that the stiffness degradation is slower with decreasing maximum stress. Alternatively, in other words, the absolute value of the slopes of the fitted stiffness degradation curves lessens with decreasing maximum stress. The data is further analyzed in Section 5.4 together with the hysteresis data.

The same procedure is used to investigate the stiffness degradation of the

90°-specimens. The overview of the fitted relative stiffness curves in Figure 5.8 shows the same behavior as for the 0°-specimens. Based on the findings for the 0°-specimens, fewer specimens were tested here overall but at more load levels. The results show that 60% of the R_{UTS} marks the limit above which the stiffness degradation is faster in the 2nd part than in the 1st part of the test. While the stiffness decreases steadily at 60% of the R_{UTS}, the curve for a maximum stress below this 60% of the R_{UTS} flattens out after the first approx. 10^3 cycles. With the parameters of the S-N curves in Table 5.2, this maximum stress level corresponds to a fatigue life of $6.4 \cdot 10^3$ load cycles, which is in the range of the transition from LCF to HCF (see Section 2.3.1).

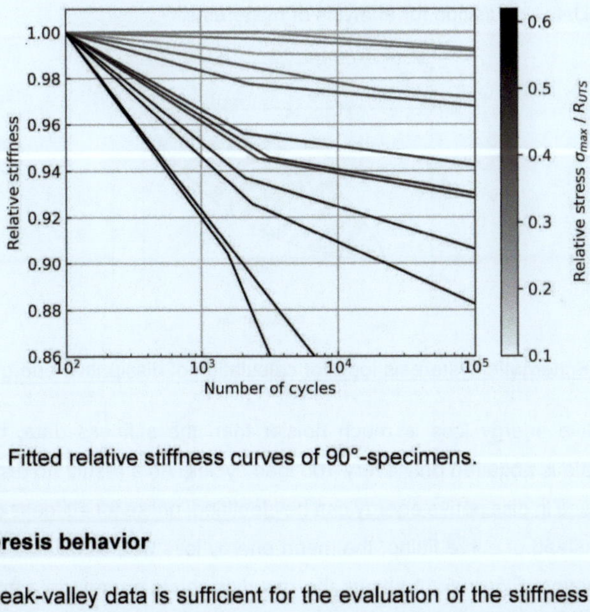

Figure 5.8: Fitted relative stiffness curves of 90°-specimens.

5.3 Hysteresis behavior

While the peak-valley data is sufficient for the evaluation of the stiffness degradation, an investigation of the hysteresis behavior requires the data of an entire loading and unloading cycle, which have to be acquired at a sufficient sample rate to map sinusoidal loading. The displacement and force of a complete cycle is sampled with 250 data points, but only every 100 cycles in order to reduce the amount of data (see Figure 5.9 for a schematic representation of the data acquisition for the analysis of hysteresis).

Subsequently, for each hysteresis loop the dissipative energy is calculated, which corresponds to the difference between the area under the loading and the unloading curves (Figure 5.10). Since the energy is lost due to internal friction or the formation thereof, this parameter is also suitable for evaluating the damage state under cyclic loading, in addition to stiffness degradation.

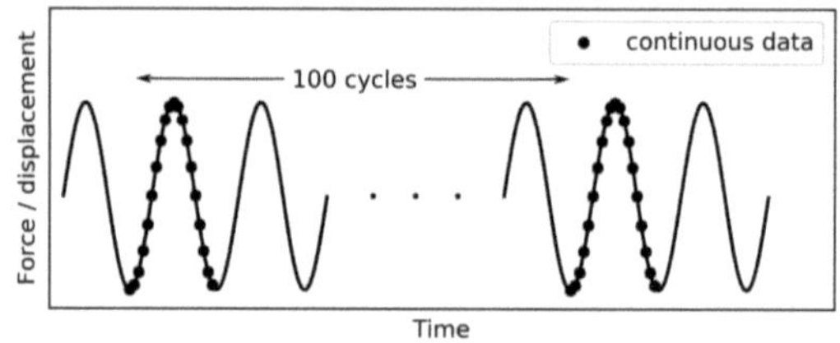

Figure 5.9: Data acquisition for analysis of hysteresis.

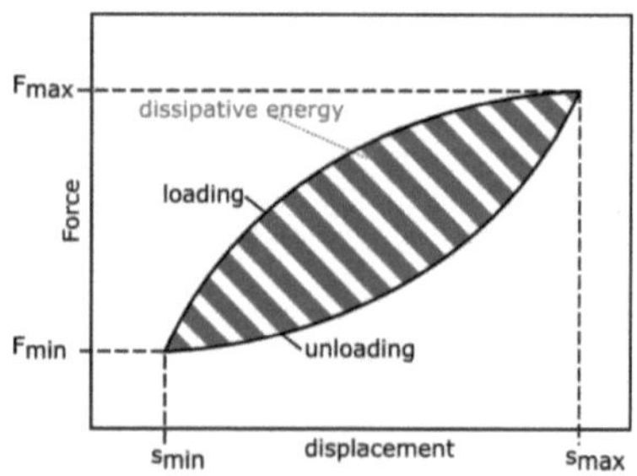

Figure 5.10: Schematic hysteresis loop for calculation of dissipative energy.

The dissipative energy loss is much noisier than the stiffness data, because the hysteresis data is acquired only every 100 load cycles. As a result, no distinct regions and no increase in dissipative energy can be identified, unlike for stiffness degradation. Therefore, instead of curve fitting, the mean energy loss was calculated and used for further evaluation. Figure 5.11 shows the raw dissipative energy data and the mean value. Since the internal damage increases with the number of cycles, the dissipative energy is expected to increase as well. This trend can be seen from the raw data, especially in the first approximately 10^3 cycles. However, since the recorded amount of data is too small, the dissipative energy is not evaluated for different ranges and only the mean value can be used for further analyses. Whether this mean value is of any significance and allows conclusions to be drawn about fatigue behavior will be presented in the following chapter.

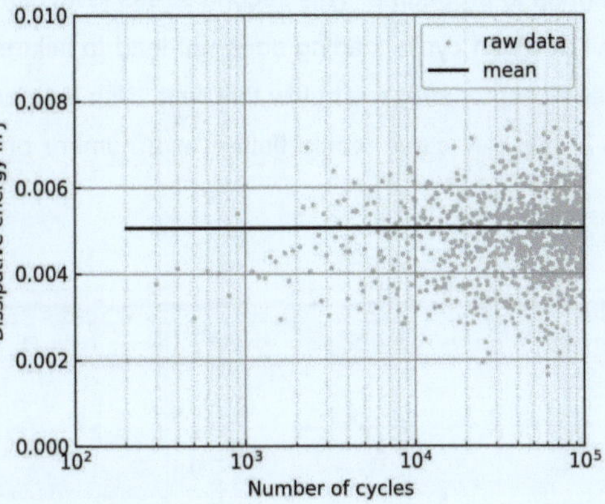

Figure 5.11: Dissipative energy data and mean value of 0°-specimen, σ_{max} = 73.2 MPa, resp. 0.48 R_{UTS}.

5.4 Discussion of fatigue behavior

In the previous two chapters, the data acquisition and processing were explained. In the following, the further evaluation is explained, which allows conclusions to be drawn about different service life ranges. For this purpose, the parameters of the stiffness degradation curves are investigated in a first step starting with the data of the specimens in the injection molding direction. Figure 5.12 shows the slope of the first part (a) and second part (b) of the stiffness degradation curve plotted against the relative maximum stress. All values were normalized to the respective maximum value at 62% of R_{UTS} for better comparability. Both slopes in Figure 5.12 are equal to zero for a maximum stress less than or equal to 18% of the R_{UTS}, which means that no stiffness degradation occurs below this stress level. This can therefore be an indication of an endurance limit below which no damage is initiated. With the parameters of the S-N curve in Table 5.2, the number of cycles to failure at this stress level can be extrapolated to approx. 10^{11}.

A second characteristic stress level, which can be deduced by the stiffness degradation data, is 26% of the R_{UTS}. For a maximum stress less than or equal to this second characteristic stress level, the stiffness decreases only in the first approx. 10^3 cycles. This can be interpreted to mean that for a maximum stress less than or equal to 26% of the R_{UTS}, some damage is initiated in the first cycles, but strain energy is insufficient for damage to propagate. This second stress level can also be interpreted as endurance limit at which cyclic loading does not lead to failure, which means that the S-N curve flattens out for stresses below this limit. With the parameters of the S-N curve in Table 5.2, the S-N curve would flatten at a number of cycles to failure of approx. $7 \cdot 10^9$.

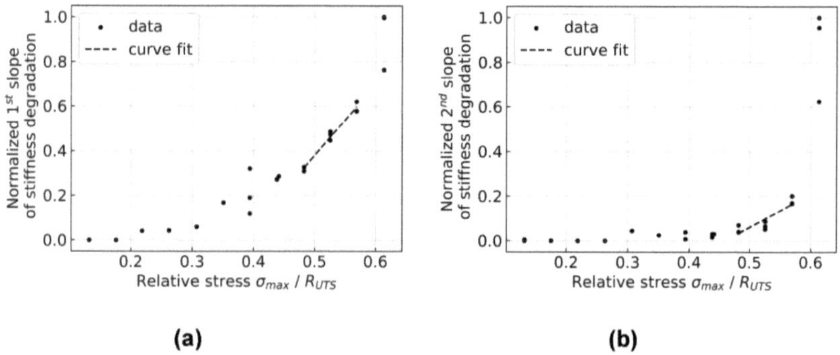

Figure 5.12: Normalized slope of stiffness degradation curves of 0°-specimens: (a) 1st slope; (b) 2nd slope.

Especially in the plot of the second slope, it can be seen without further analysis steps that the data can be divided into three ranges: in the first range, below a certain maximum stress, there is almost no decrease in stiffness; it increases linearly in the second range until it increases disproportionately in the third range. The second range, in which the stiffness degradation depends approximately linearly on the maximum stress, thus subdivides these three domains. It was therefore investigated in which range the data sets can best be approximated by a linear equation to identify the second range. For this purpose, the curve parameters were determined using *python's SciPy* library so that the standard error is minimized. Figure 5.12 shows the range in which the slopes of the stiffness degradation can best be represented by a linear curve. For both slopes these limits result in 48% and 57% of the R_{UTS}. Between 48% and 57% of the R_{UTS}, the stiffness degradation thus depends nearly linearly on the maximum stress. Below 48% of the R_{UTS}, the stiffness degradation proceeds much more slowly, while above 57% of the R_{UTS} the stiffness degradation increases disproportionately. The later stress level is the same as the stress level determined in Figure 5.7, above which the stiffness degradation in the second part of the test progresses faster than in the first cycles.

In a second step, these stress bounds are correlated to the fatigue lives with the parameters of the S-N curve listed in Table 5.2. The lower bound corresponds to a fatigue life of 1.4·10^6, whereas the upper bound corresponds to a fatigue life of 4.5·10^4. With the definition in Section 2.3.1 of the different fatigue ranges, such as LCF, HCF, and VHCF, these number of load cycles limit the range of HCF. Since stiffness degradation is directly related to the damage mechanisms in the material, this phenomenological observation leads to the assumption that either the underlying damage mechanisms or their propagation rates change at the transition from LCF to HCF and from HCF to VHCF.

In the next step, the presented procedure to distinguish the region of HCF by analyzing the stiffness degradation data is verified with the data of the tests transverse to the direction of injection molding. For the stiffness degradation transverse to the direction of injection molding, less test data, but at smaller maximum stress intervals in the beforehand identified regions of interest, is available.

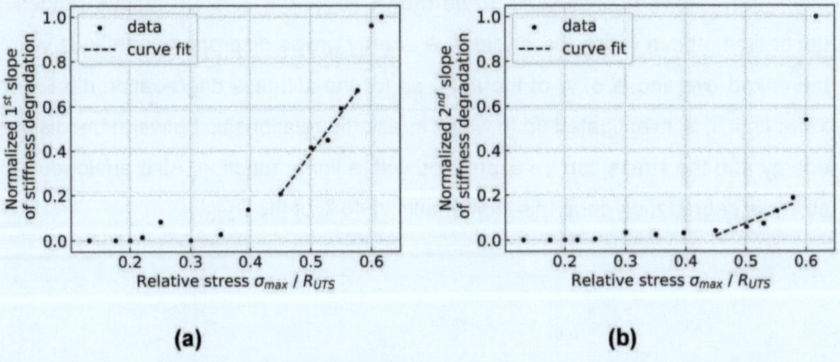

Figure 5.13: Linear curve fit to stiffness degradation data of 90°-specimens: (a) 1st slope; (b) 2nd slope.

Figure 5.13 shows the slopes of the stiffness degradation curves of the tests transverse to the direction of injection molding. Up to 20% of R_{UTS} no stiffness degradation is observed. At 22% of R_{UTS} a stiffness degradation is measured only in the first cycles, which does not progress in the second part of the test. With the parameters in Table 5.2 for the 90°-specimens, it can be assumed that the S-N curve flattens later than between 7·10^9 and 1.4·10^{10} load cycles. Again, it is visually apparent that the data can be divided into three regions where the stiffness degrades at different rates depending on the maximum applied stress. The stress boundaries separating these areas are determined to be 45% and 57% of the R_{UTS} corresponding to fatigue lives of 1.6·10^6 and 1.9·10^4 load cycles. This shows that the different fatigue life ranges can also be estimated with the stiffness degradation data from the tests transverse to the injection molding direction.

In the rest of this section, the data of the mean dissipated energy are evaluated and analyzed analogously to the stiffness degradation data. Figure 5.14 shows the normalized dissipative energy data plotted against the relative maximum stress. It becomes apparent, that the dissipative energy does not vanish even for the smallest maximum stress in this test series. Since dissipative energy emerges from the material due to internal friction or damage, this suggests, in contradiction to the observations of stiffness degradation, that there is no fatigue limit above the investigated stress of 13% of R_{UTS}. However, this observation and interpretation must be considered against the background of the large scatter of dissipative energy data.

As for the stiffness degradation data, the dissipative energy data were examined to see if they allowed conclusions to be drawn about the different fatigue ranges. The upper limit, above which the dissipative energy grows disproportionately, is visible to the naked eye and is 57% of the R_{UTS}, as for the stiffness degradation data. For the lower limit, it is investigated up to which stress the relationship between the dissipative energy and the stress can be expressed with a linear function. Also analogous to the stiffness degradation data, this limit results in 48% of the R_{UTS}.

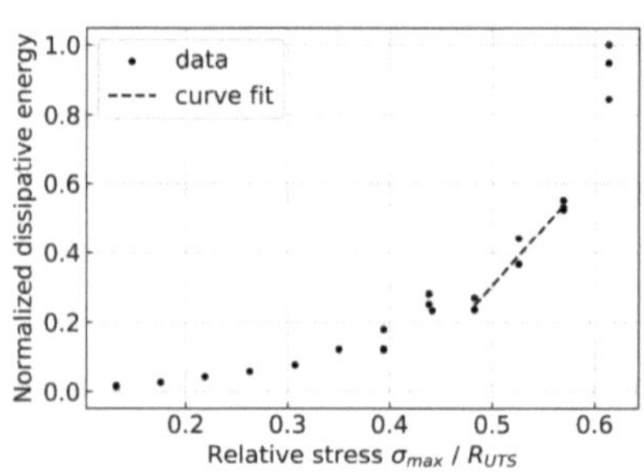

Figure 5.14: Dissipative energy data of 0°-specimens with linear curve fit.

Figure 5.15 shows the dissipative energy data of the 90°-specimens for verification of the above-discussed findings. For the test data transverse to the direction of injection molding, it can also be seen that the dissipative energy does not disappear completely. Moreover, the stress levels identified are again consistent with those found through the analysis of the stiffness degradation data. Thus, using the dissipative energy data, the range of the HCF can also be separated from the range of the LCF and the VHCF.

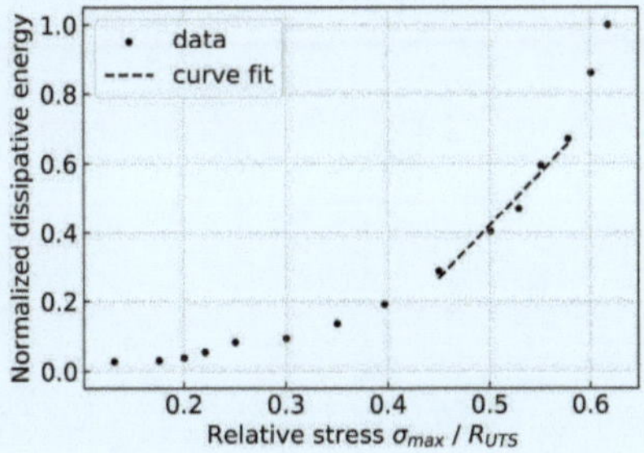

Figure 5.15: Dissipative energy data of 90°-specimens with linear curve fit.

6 Acoustic emission testing

Section 2.3.3 explains the AE testing procedure and its application to SFRT. It becomes clear that due to the large amount of data, AE is mostly used for quasi-static tests and not for monitoring complete fatigue test series. To address this challenge, this work compares the acoustic events in quasi-static tests on previously non-preloaded specimens with the acoustic events in residual strength tests on preloaded specimens. For this purpose, some of the specimens were subjected to cyclic preloading, while the rest remained non-preloaded. For the non-preloaded specimens, the AE can capture the damage caused by the quasi-static tensile loading. However, for the preloaded specimens, AE can only capture the damage not already caused by the cyclic preloading. Therefore, by comparing the AE data of non-preloaded and preloaded specimens, the damage condition of the preloaded specimens can be evaluated. The following chapters list the fatigue preloading, the test set-up of the quasi-static tests using AE, and presents and interprets the acquired test data.

6.1 Test set-up

6.1.1 Fatigue preloading

The fatigue preloading was conducted on the same testing machine as the tests described in Section 5 using the same parameters (R = 0.1, f = 10 Hz, 23 °C and 40% relative humidity). Likewise as in Section 5, the study focusses on specimens in the direction of injection molding and subsequently uses specimens transverse to the direction of injection molding to verify the findings.

With regard to the fatigue preloading, two different test series can be distinguished: while the first series of tests varies the maximum stress level, in the second series the maximum stress level is kept constant and the number of load cycles is varied. Matching the conclusion from the analysis of the stiffness degradation data (Section 5.2) that the stiffness degrades at a constant rate after 10^5 cycles, the specimens of the first series are preloaded with 10^5 load cycles. The maximum stress during the preloading varies between 13% and 53% of R_{UTS}.

In the second series, the specimens are preloaded with a maximum stress of 40% of R_{UTS} up to a number of load cycles ranging between 10^2 and 10^7. Table 6.1 shows the test matrix for the preloading of the specimens in the direction of molding.

Table 6.1 Test matrix of fatigue preloading of 0°-specimens

Maximum stress in MPa	Relative maximum stress	Number of cycles					
		10^2	10^3	10^4	10^5	10^6	10^7
79.8	0.53				x		
75.5	0.50				x		
73.2	0.48				x		
66.5	0.44				x		
59.8	0.40	x	x	x	x	x	x
53.2	0.35				x		
46.6	0.31				x		
39.9	0.26				x		
33.2	0.22				x		
30.2	0.20				x		
26.6	0.18				x		
20.0	0.13				x		

Table 6.2 Test matrix of fatigue preloading of 90°-specimens

Maximum stress in MPa	Relative maximum stress	Number of cycles
57.2	0.50	10^5
51.4	0.45	10^5
45.3	0.40	10^5
40.0	0.35	10^5
34.3	0.30	10^5
28.6	0.25	10^5
25.2	0.22	10^5
22.9	0.20	10^5
20.1	0.18	10^5
15.1	0.13	10^5

Due to the findings presented later for the specimens in the injection molding direction, the number of cycles for preloading the specimens transverse to the injection molding direction is kept constant and only the maximum stress is varied (see Table 6.2).

6.1.2 Quasi-static tests with AE

Quasi-static tensile tests are performed using AE analysis and a digital image correlation (DIC) system. The AE is used to identify individual damage events that occur during the tensile loading. Correlation with DIC can be used to determine the global and local strain on the specimen surface at the level at which individual damage events occur.

The tests are run on a mechanical testing machine (RetroLine, ZwickRoell, Ulm, Germany) at room temperature (23°C and 40% relative humidity) under displacement control. According to DIN EN ISO 527-1 [97], the speed of movement of the machine crosshead is set to 1 mm/min. The force is measured with a 10 kN load cell. Pneumatic jaws at a distance of 60 mm clamp the specimens. The local strain is measured by means of DIC using the camera system Aramis4M (GOM GmbH, Braunschweig, Germany). The speckle pattern is compared to the reference pattern provided by Gom GmbH for a measuring area of 65 x 52 mm^2. Prior to each test, a reference picture is taken and analyzed to determine whether the speckle pattern allows continuous evaluation (see Table 6.3 for the parameters of the DIC).

Table 6.3 Parameters of DIC

Parameter	Value
Focal length	100 mm
Resolution	2358x1728 Pixel
Sample rate	5 Hz
Speckle size	16 Pixel
Speckle distance	12 Pixel

To capture the acoustic activity during tensile loading, two sensors are mounted symmetrically at a distance of approx. 36 mm (see Figure 6.1). Since the sensor position has a significant influence on the quality of the localization of sound events, the sensor positions are measured in a first step with the help of a ruler prior to each test. In a second step, the measured distance between the two sensors is verified by three pencil lead breaks at the height of each sensor. With the mean difference in the arrival times for the pencil lead breaks and the sound velocity the real distance is approximated. However, this allows only the determination of the distance between the sensors and not the absolute position. Therefore, a deviation determined from the measured distance is divided in half between the two sensor positions. In this way, a higher accuracy can be achieved for the sensor positions than by purely manual determination. The sound velocity has been determined on the injection-molded plates in different angles to the injection molding direction as listed in Table 6.4 [104].

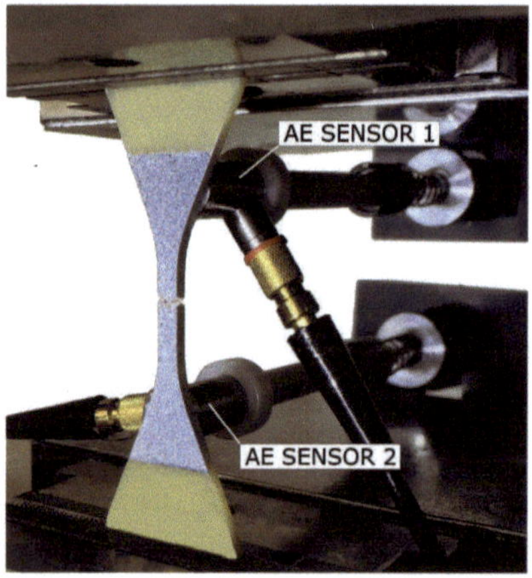

Figure 6.1: Test set-up with AE sensors and speckle pattern for DIC.

Since the different damage mechanisms of injection molding material cause signals in different frequency ranges, broadband sensors were used. The selected sensors AE104A (Vallen Systeme GmbH, Wolfratshausen, Germany) are characterized by their small aperture of 8 mm and their flat frequency response between 100 and 400 kHz.

Table 6.4 Sound velocity depending on the angle to the injection molding direction [104]

Angle to the injection molding direction	Velocity in m/s	Standard deviation in m/s
0°	3274	125
90°	2759	112

Since broadband sensors are susceptible to the detection of background or machine noise, it must be ensured that the sensor is mounted at a sufficient distance from the fixture and is fully coupled to the specimen. A medium viscosity silicone (Korasilon, Kurt Obermeier GmbH & Co. KG; Bad-Berleburg, Germany) is used to couple the sensors to the specimen. To check that the sensors are fully and evenly coupled, a pencil lead is broken centrally between the two sensors and the measured signals at the two sensors are compared. The use of the same lead thickness, hardness and length ensures the comparability of the results for different tests.

The signals are recorded with the system AMSY−5 (Vallen Systeme GmbH, Wolfratshausen, Germany) and with the use of two preamplifiers (AEP4, Vallen Systeme GmbH, Wolfratshausen, Germany) with the gain set to 40 dB. Before each test, a recording of the machine and background noises is performed in order to verify the coupling of the sensors to the specimen. The threshold is set at $2.52 \cdot 10^{-5}$ mV which corresponds to 28 dB. As soon as the threshold is exceeded, the respective sensor starts to store the data with a sample rate of 10 MHz. The total length of a signal is defined to 4096 samples, resulting in a hit duration of 409.6 µs. A so-called pre-trigger, i.e. the data before the first registered crossing of the threshold, is set to 512 samples. In the following, each recorded signal is called a *hit*.

The software stores the data in two different formats: the so-called transient database consists of the waveforms of each individual hit. In the so-called primary database, predefined parameters of the hits such as the signal arrival time, the maximum amplitude, or the signal duration are stored. A transient index is assigned to each recorded hit, which can be used to uniquely link the two databases. It is not possible to trigger the AE system with a trigger signal from the mechanical testing machine, but it is possible to measure parametric inputs such as the load or displacement signal from the mechanical testing machine. With the help of these parametric inputs, the acoustic data can be superimposed with the stress-strain data by correlating the time of the two data acquisitions. Therefore, a customized *python* code is used, which is described in the following.

6.2 Evaluation method

The evaluation algorithm is visualized in Figure 6.2 and is composed of the five blocks:

1. Read data
2. Merge data
3. Filter AE data
4. Analyze AE data
5. Export data.

6.2.1 Read data

The first block consists of reading the data. In a first step, the configuration data is read, which must be defined for each test individually. It consists of set-up parameters such as the sensor positions, the sound velocity, the threshold, and the window for the determination of the arrival time with the AIC (see eq. 2.3). The test matrix with the specimen information such as specimen cross section and preloading conditions is also stored in the configuration data. In a second step, the test data is read in consisting of the AE data and the DIC data.

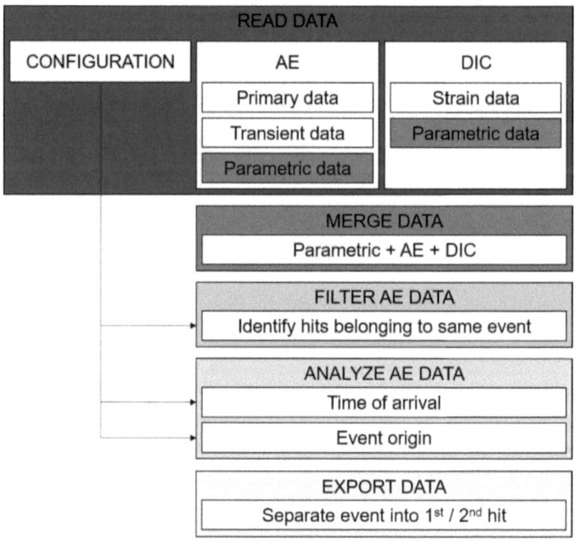

Figure 6.2: Evaluation algorithm implemented with *python*.

6.2.2 Merge data

In the second section, the data are reduced and superimposed in time. For this purpose, the start and end of the experiment are defined and the excess data is deleted. Since the machine data are transferred to the AE and DIC measuring systems via digital inputs, they are contained in both data sets as so-called parametric data. This parametric data is therefore used to merge the acoustic data with the strain data.

6.2.3 Filter AE data

Subsequently, the AE data is reduced by identifying events that occurred between the two sensors. For this purpose, the maximum time span that may lie between two successive hits on the two sensors is calculated from the sound velocity and the sensor distance. This is compared with the hit arrival times from the primary database. All other hits that were only registered by one sensor within the calculated maximum time span are then deleted. This section of the program is referred to as the event finder and is run a second time afterwards with the corrected arrival times. For the first filter step with the event finder, the maximum time span between two successive hits was set larger to avoid accidentally sorting out valid events.

6.2.4 Analyze AE data

The timestamp of the hits in the primary database is the time of the first threshold crossing registered by the system. This method of determining the time of arrival is simple on the one hand, but inaccurate on the other. Numerous alternative methods have been developed in earthquake research aiming at a more accurate determination of the time of arrival (see Section 2.3.3). In the present work, the best results compared to a manual determination of the arrival time are obtained by applying the AIC. In the context of sound velocity determination, signal arrival times were compared using different calculation methods [104]. Figure 6.3 visualizes the arrival times of a signal determined by the first threshold crossing and by the AIC and thereby illustrates the greater accuracy of the second criterion.

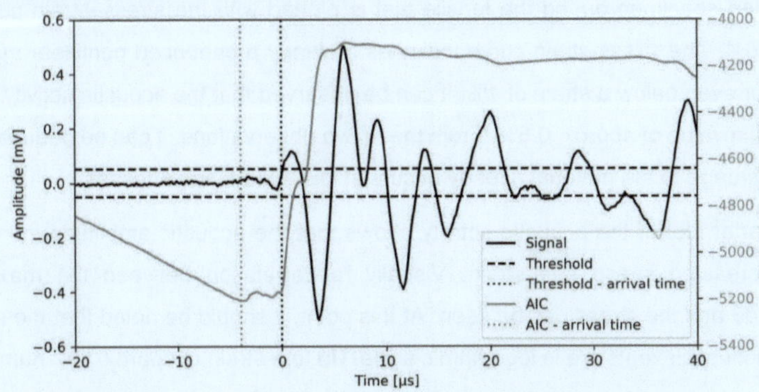

Figure 6.3: Signal and arrival time determined by first threshold crossing and AIC.

With these adjusted arrival times, the event finder runs again to identify hits that do not constitute an event in order to filter them out of the database. Based on the corrected arrival time, the origin and the occurrence time are determined for the remaining events. The use of two sensors only allows one-dimensional location on a straight line connecting the two sensors. The small distance between the two sensors requires a high transient resolution of the signals. To reduce the data further, only the events that occurred within the gauge length are filtered.

6.2.5 Export data

In a final filtering step, the events were divided into a first and a second hit, with the first hit denoting the one first detected by the sensor closest to the origin of the events. Only the first hits were included in the subsequent analysis.

The principle of acoustic emission testing is that each measured acoustic event is associated with a damage event. Therefore, the analysis of the accumulated acoustic events allows to draw conclusions about the damage state. In the following sections, the acoustic activity of the specimens in the injection molding direction is presented first and subsequently reviewed for those transverse to the injection molding direction.

6.3 Specimens in the injection molding direction

6.3.1 Quasi-static tests

A typical acoustic activity with the accumulated number of acoustic hits of a non-preloaded specimen during the tensile test is plotted with the stress–strain curve in Figure 6.4. The stress-strain curve indicates a clearly pronounced nonlinear material behavior even below a strain of 1%. It can be observed that the acoustic activity starts at a strain value of approx. 0.5%. From these two observations, it can be deduced that initial damage to the material already occurs at low strains and stresses.

The scatter plot of the acoustic activity shows that the acoustic amplitude increases with increasing stress and strain. Visually, a correlation between the maximum amplitude and the stress can be seen. At this point, it should be noted that the scales of the acoustic events are in logarithmic scale. Up to a strain of approx. 3%, numerous events with an acoustic amplitude below 1 mV occur. After that, these low-amplitude events occur only sporadically, and the acoustic amplitude of most events is above

1 mV. This can be interpreted as different damage mechanisms prevailing in the first part of the experiment compared to the second part of the load.

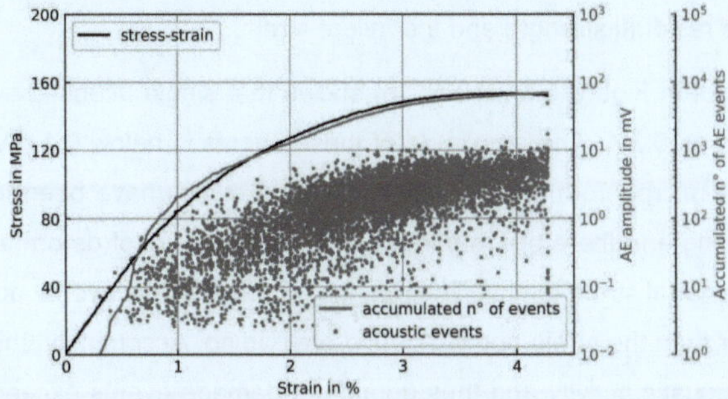

Figure 6.4: Stress–strain curve, acoustic amplitude and accumulated number of acoustic hits of non-preloaded 0°-specimen.

It is also apparent that above approx. 2% strain, which corresponds to a stress of approx. 130 MPa, the point cloud is much denser than below. This means that qualitatively more damage events occur at higher strains and stresses than at lower strains and stresses. This is also evident when plotting the cumulative number of AE events on a linear scale. However, the logarithmic representation method has other advantages, which become clear when comparing the non-preloaded with the preloaded specimens. On a logarithmic scale, the curve of the accumulated number of acoustic events initially shows a steeper course that flattens out in the course of the experiment. From about 1.5% strain, which corresponds to about 110 MPa stress, the stress-strain curve and the curve of the accumulated number of acoustic events on logarithmic scale run parallel.

6.3.2 Residual strength tests for preloading with different maximum stresses

In the following, the acoustic response in the residual strength tests of the specimens preloaded with different maximum stresses is presented. Figure 6.5 shows the stress–strain curves, the scatter plot of the maximum acoustic amplitude and the accumulated number of acoustic events of three specimens preloaded with the same number of load cycles of 10^5 but each with a different maximum stress level. The maximum stress levels are 33.3 MPa, 53.2 MPa and 79.8 MPa corresponding to 22%, 35% and 53% of R_{UTS} and are indicated in the plots with a dotted line.

In the residual strength test of the specimen with a cyclic preloading with 79.8 MPa (= 53% of R_{UTS}) the specimen failure already occurs at a strain of 3.5%. For the specimens preloaded at lower stresses, the cyclic preloading shows no significant impact on the residual strength and the failure strain.

The scatter plot in Figure 6.5 (a) and (b) shows that single acoustic events occur at strains as low as 0.2%. The amplitude of these events is below 0.1 mV. This can be interpreted as microdeformation at existing cracks, which have been initiated by the cyclic preloading and therefore require only a small amount of deformation energy to grow in the residual strength test. The majority of the acoustic events occur at a strain slightly higher than the strain applied during preloading. Accordingly, the stress at the start of the acoustic activity and thus during the damage events exceeds the applied maximum stress of the fatigue loading. These findings are contradictory to the Felicity effect, that first became apparent during the investigation of tanks and vessels made of glass fiber reinforced plastics [98]. The Felicity effect is observed in the load increase test, where the acoustic activity already starts at stress levels below the previously applied stress. The difference to the residual strength tests is the applied number of load cycles. The offset between the strain and stress applied during the preloading and the strain and stress at which the acoustic activity starts indicates that cyclic loading causes more damage than quasi-static loading.

All three scatter plots in Figure 6.5 show an increasing maximum amplitude in the course of the test. Up to about 3% strain, the envelope of the scatter plot is approximately linear, provided the amplitude is plotted on a logarithmic scale. From approx. 3% strain, the envelope flattens out and the maximum amplitude of the sound events no longer increases. See Figure 6.5 (c) for a visualization of the above-described observations on the envelope of the maximum amplitude.

The comparison of Figure 6.4 and Figure 6.5 reveals that the point clouds and the accumulated number of acoustic events of the preloaded specimens differ significantly from those of the non-preloaded specimens. The preloaded specimens show fewer acoustic events with low amplitudes than the non-preloaded specimen does. This indicates that the damage events corresponding to acoustic events with low amplitudes already occurred during cyclic preloading.

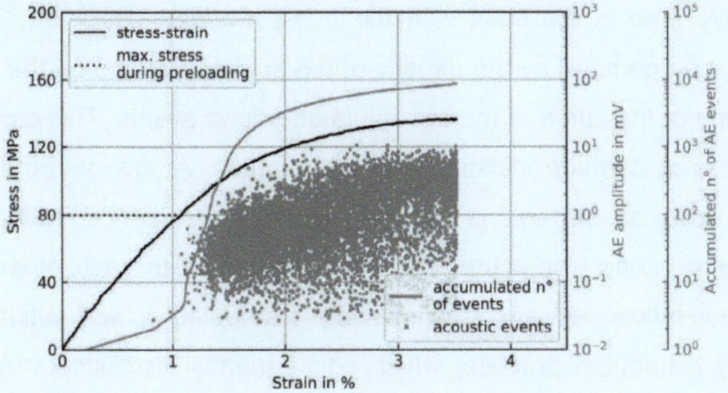

(a)

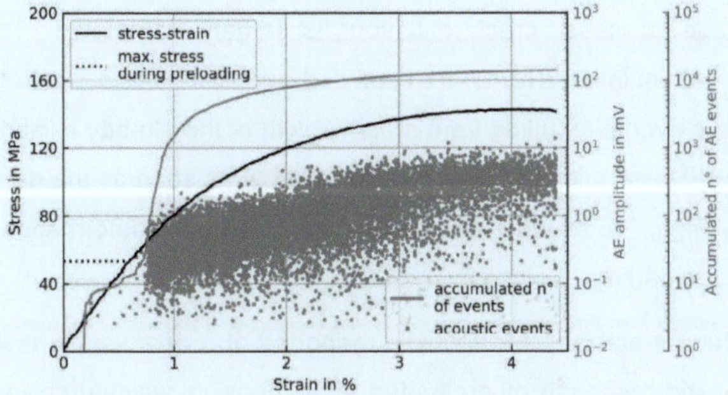

(b)

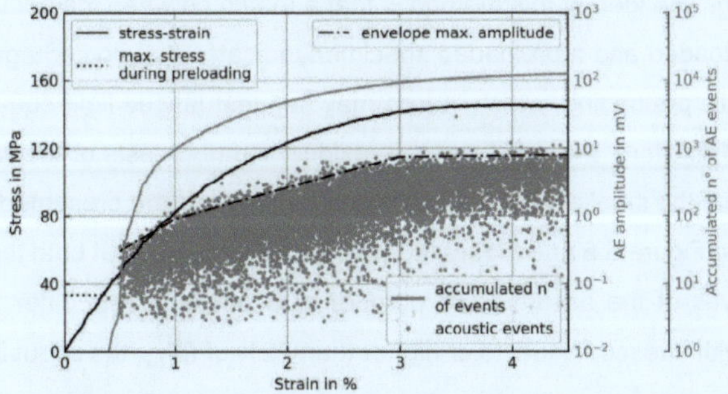

(c)

Figure 6.5: Stress–strain curve, acoustic amplitude and accumulated number of acoustic hits of preloaded 0°-specimens: σ_{max} = (a) 53%; (b) 35%; (c) 22% R_{UTS}.

The acoustic events in the residual strength tests with the preloaded specimens start more promptly than in the tests with the non-preloaded specimen. This becomes apparent on the one hand by the density of the scatter plot and on the other hand by the initial slope of the curve of the accumulated sound events. This can be explained by the process of damage initiation and propagation. At low deformations, the first microcracks occur at the regions with the highest stress concentrations, such as at fiber ends or in areas where the fibers are very close to each other. Since these microstructural influences are stochastically distributed, crack initiation is also a stochastically distributed process, which consequently also occurs in a distributed manner during loading. After crack initiation, these microcracks continue to grow as soon as the deformation work is large enough. During cyclic preloading, the first part of the damage process, crack initiation, is largely completed. Therefore, only occasional crack initiation events are recorded during the residual strength tests and mainly acoustic events resulting from crack growth of the already existing microcracks are registered. These microcracks continue to grow as soon as the deformation work is sufficiently high, i.e. at the earliest after exceeding the deformation work applied during cyclic preloading.

In the previous paragraph, the acoustic response of a quasi-static test is compared with the acoustic response of preloaded specimens in residual strength tests. The differences of the acoustic activity clarified that damage had been introduced by the cyclic loading. The idea of this method is that a match between the acoustic response of a non-preloaded and a preloaded specimen indicates that no damage has resulted from the cyclic preloading, which in turn may suggest fatigue limit equal to the cyclic preloading. Therefore, the results of the residual strength tests of the two specimens preloaded with the smallest stresses in this series of tests are presented in Figure 6.6. By comparing Figure 6.5 and Figure 6.6, it becomes evident, that both the scatter plots and the curves of the accumulated number of acoustic events differ. While for the preloading with stresses equal to or higher than 22% of R_{UTS}, the acoustic events start promptly after exceeding a certain strain or stress, the scatter plots of the specimens with a preloading of equal to or lower than 18% of R_{UTS} show a lower density at the beginning of the test. A comparison of Figure 6.4 and Figure 6.6, on the other hand, suggests a match in acoustic activity.

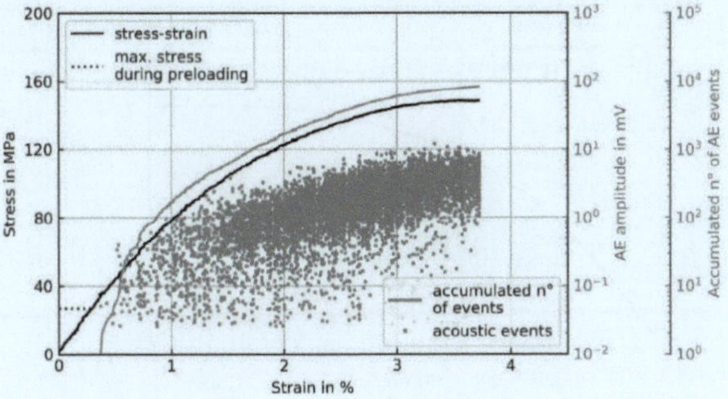

(a)

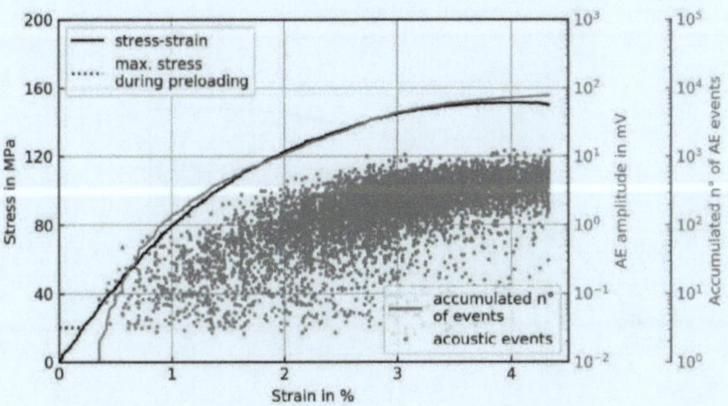

(b)

Figure 6.6: Stress–strain curve, acoustic amplitude and accumulated number of acoustic hits of preloaded 0°-specimens: σ_{max} = (a) 18%; (b) 13% R_{UTS}.

6.3.3 Residual strength tests for preloading with different number of load cycles

After having assessed the influence of different maximum stress levels during the cyclic preloading on the acoustic activity in residual strength tests, this chapter investigates the influence of different numbers of load cycles on the AE in residual strength tests. Figure 6.7 shows the stress–strain curve, the amplitude, and the accumulated number of hits of the 0°-specimens preloaded with the same maximum stress of 40% of R_{UTS} (59.8 MPa) but with a different number of cycles. Representative test data are shown for a preloading with 10^6, 10^4 and 10^2 cycles. The figures for 10^7, 10^5 and 10^3 load cycles can be found in Appendix A.

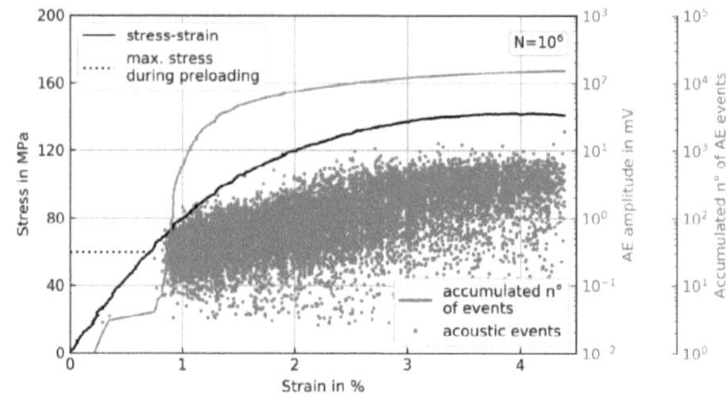

(a)

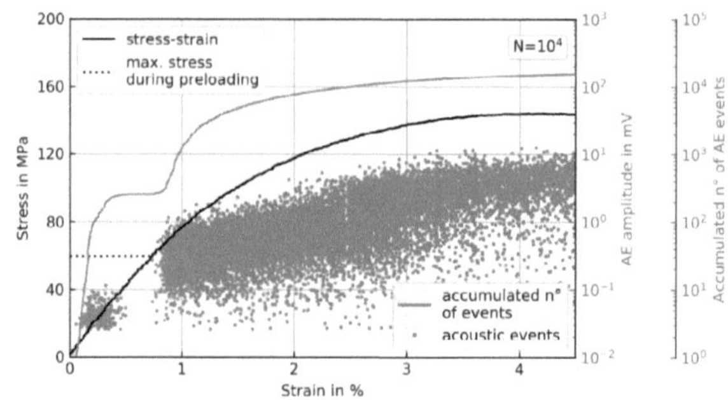

(b)

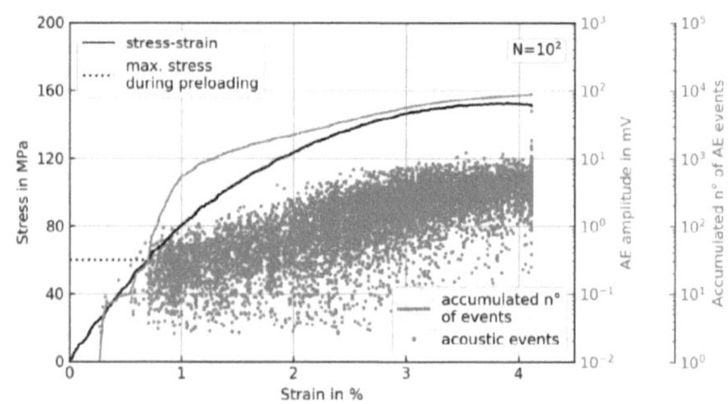

(c)

Figure 6.7: Stress–strain curve, acoustic amplitude and accumulated number of acoustic hits of 0°-specimens preloaded with σ_{max} = 40% R_{UTS}: (a) 10^6; (b) 10^4; (c) 10^2 load cycles.

AE data during the residual strength test of specimens preloaded with 10^5, 10^6 and 10^7

load cycles show the same behavior. In the test with a preloading of 10^4 cycles, a few hundred acoustic hits with low amplitude occur at low strain. After these first acoustic events, there is a pause until the next events occur at a strain of about 0.8%. This observation can be explained with the damage process under cyclic loading: in the first part of the lifetime, micro-cracks are initiated up to a certain number of load cycles. This is followed by a phase of stable crack growth. The single AE events occurring after 10^4 but not after 10^5 load cycles indicate that the phase of stable crack growth starts between 10^4 and 10^5 load cycles, at least for a cyclic loading with a maximum stress of 40% of R_{UTS}. Before reaching this state, further micro-cracks are induced that require little additional deformation work in the residual strength tests to grow and therefore occur at low strains. Figure 6.7 shows, that the acoustic fingerprint of the residual strength tests on the specimens preloaded with 10^2 load cycles resembles that of the quasi-static tests except for several isolated acoustic events at low strain.

6.4 Discussion of acoustic response

The previous chapter presents the results of the quasi-static test of a non-preloaded specimen and the residual strength tests on specimens with different preloading. Since it is difficult to superimpose the scatter plots of the different experiments to directly compare the acoustic activity, the accumulated number of acoustic events are compared. Figure 6.8 shows the accumulated number of acoustic events of the non-preloaded and the preloaded specimens and visualizes the above presented reflections.

Direct comparison demonstrates that the curves of the preloaded specimens at the beginning of the acoustic activity are steeper than those of the non-preloaded specimens. Figure 6.8 (b) reveals that the characteristics of the accumulated number of hits of the preloaded specimens change at a maximum stress during preloading below 26% of the R_{UTS} and approach the curves of the non-preloaded specimens. This observation suggests that damage initiation at this stress level was not yet fully completed by the cyclic preloading and that the damage behavior changes below this stress level.

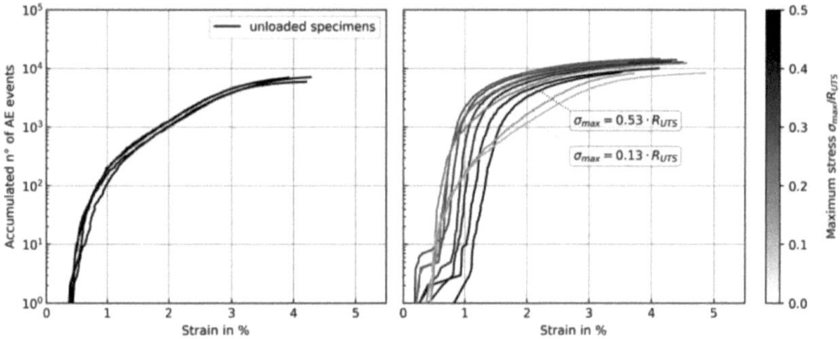

(a) (b)

Figure 6.8: Accumulated number of acoustic events of 0°-specimens: (a) non-preloaded; (b) preloaded sorted by maximum relative stress level during preloading.

For better visual comparison, Figure 6.9 shows the accumulated number of AE events of the non-preloaded specimens and the specimens preloaded with the four lowest maximum stresses. The graph reveals that the acoustic activity at a maximum stress during preloading less than or equal to 18% of the R_{UTS} behaves similarly to that of the non-preloaded specimens. This suggests that the applied cyclic load during this preloading did not cause any measurable damage. Accordingly, the AE results of the residual strength tests indicate that for the investigated material PA6.6GF40 in the injection molding direction, a fatigue limit exists at the earliest at maximum stresses less than or equal to 18% of the R_{UTS} or, with respect to the number of load cycles, for more than 10^{11} load cycles. This is consistent with the results of the stiffness degradation studies in Section 5.4. Likewise, stiffness degradation was measured here for a maximum stress greater than 18% of R_{UTS}, indicating that no fatigue limit exists above this stress level.

Figure 6.9 also shows that the total number of AE events for the specimens preloaded with a maximum stress of 20% and 22% of R_{UTS} significantly exceeds the total number of AE events for the undamaged specimens and the specimens with maximum stress of 13% and 18% of R_{UTS}. This observation indicates that the damage process under cyclic loading is different from that under quasi-static loading. The higher number of AE events can be interpreted to mean that cyclic loading initiates damage at more sites than quasi-static loading, but the cyclic stress is not sufficient for this damage to propagate. These cracks start to grow as soon as the stress is high enough, which is why they are registered as AE events in the residual strength tests.

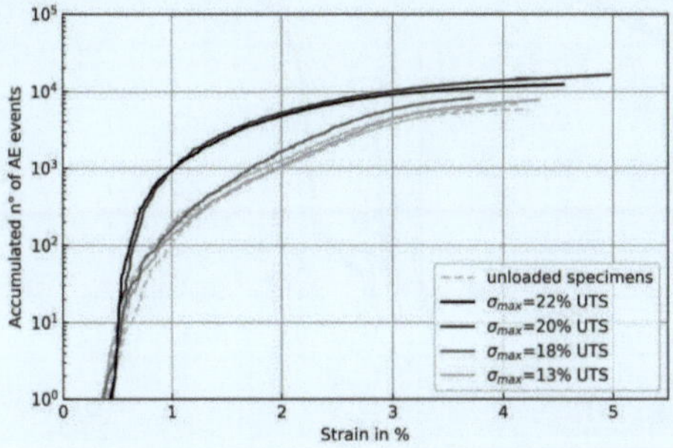

Figure 6.9: Accumulated number of acoustic events of 0°-specimens, non-preloaded and preloaded with σ_{max} = 22%, 20%, 18%, 13% R_{UTS}.

The next step is to analyze the results from the tests with the same maximum stress during the cyclic preloading but with a different number of load cycles. Figure 6.10 summarizes the accumulated number of AE events for the undamaged specimens and the specimens that were preloaded with 40% of R_{UTS} and the number of load cycles ranging between 10^2 and 10^7. The grey value indicates the number of load cycles respectively the percent of lifetime N/N_f. The figure illustrates, that the acoustic response of the preloaded specimens differs from the AE activity of the non-preloaded specimens. In Figure 6.10 (b), it is noticeable that the curve of the specimen preloaded with 10^2 load cycles runs flatter than the other curves. While a few hits occur at a low strain for a preloading with 10^3 and 10^4 load cycles, the three specimens preloaded with 10^5, 10^6 and 10^7 load cycles show the same trend of accumulated number of acoustic events.

The above observations lead to the conclusion that fatigue loading damage occurs in three stages. For a preloading with 40% of R_{UTS}, these three stages are as follows:

- Stage 1: $N < 10^3$;
- Stage 2: $10^3 \leq N < 10^5$
- Stage 3: $N \geq 10^5$.

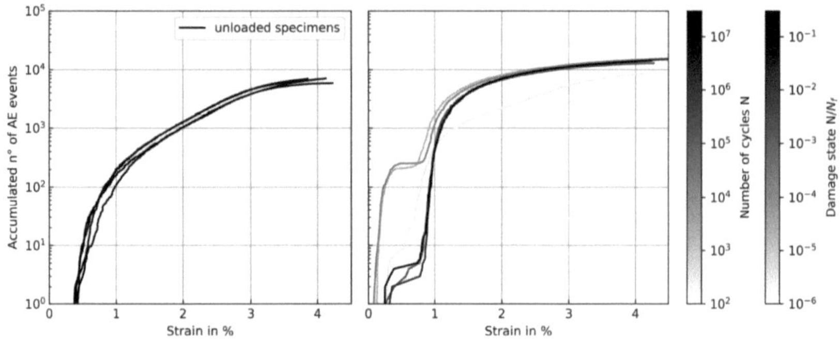

(a) (b)

Figure 6.10: Accumulated number of acoustic events of 0°-specimens: (a) non-preloaded; (b) preloaded with σ_{max} = 40% R_{UTS} and sorted by upper number of load cycles during preloading.

In the first stage, only a few damage events are measured at the beginning of the residual strength test. This suggests that fatigue loading with a number of load cycles of less than 10^3 cycles did not cause much damage. The second phase is characterized by the occurrence of a group of acoustic events with low amplitude at a small strain. This indicates that micro-cracks develop between 10^3 and 10^5 load cycles, which only require low deformation energy to grow further in the residual strength tests. The AE behavior in the residual strength tests changes again if a preloading was applied up to the third stage, indicating that a steady damage state is reached after 10^5 load cycles. These findings are also consistent with the stiffness degradation and dissipative energy results in Section 5.4. Data analysis of stiffness degradation and dissipative energy also indicate the different fatigue ranges.

To verify the results and conclusions presented above, they are compared in the following to specimens transverse to the direction of injection molding. The test matrix during the preloading of the 90°-specimens consists of different maximum stress levels with a constant number of load cycles of 10^5 with the aim of verifying the fatigue limit estimation method. Figure 6.11 shows the acoustic activity of a non-preloaded specimen transverse to the injection molding direction. Qualitatively, the AE activity looks like that of the specimens in the injection molding direction in Figure 6.4: low-amplitude acoustic events below 1 mV occur in the first third of the test, which are hardly recorded in the last third of the test. By the last third of the test, both the amplitude and the number of events increase significantly.

Quantitatively, differences between the tests in and transverse to the injection molding direction become apparent, especially when analyzing the accumulated number of AE events. For the 0°-test, the number of accumulated AE events increases faster than for the 90°-test. Thus, the first 10^2 AE events are already reached at approx. 0.88%±0.06% strain for the 0°-specimen and only at approx. 1.10%±0.05% strain for the 90°-specimen. Analogously for the first 10^3 AE events: these are reached at approx. 1.90%±0.07% strain for the 0°-specimen and at approx. 2.30%±0.02% strain for the 90°-specimen. Not only the number of accumulated acoustic events are lower for the 90°-specimen than for the 0°-specimen, but also the envelope of the maximum amplitude. While the maximum amplitude in the test with the 90°-specimen is approx. 7 mV, in the tests with the 0°-specimen it is approx. 10 mV. These observations can be explained by the fiber orientation tensor, which shows that more fibers are oriented in the injection molding direction than transverse to the injection molding direction. Therefore, the 0° specimens have a higher stiffness and thus a higher stress state at the same strain than the 90° specimens. This results in higher energy stored in the material, which is released by damage thus generating acoustic events with greater energy release.

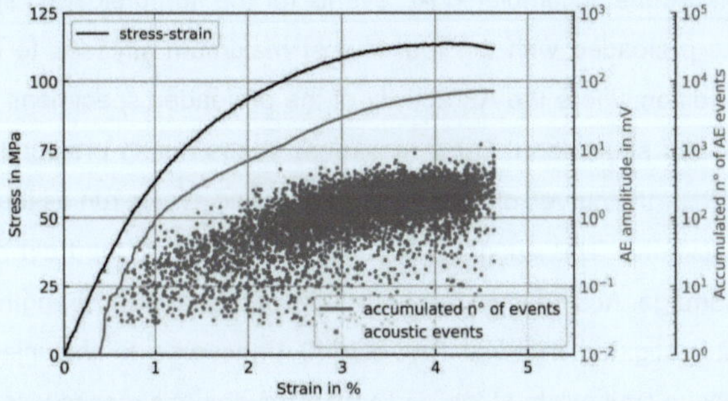

Figure 6.11: Stress–strain curve, acoustic amplitude and accumulated number of acoustic hits of non-preloaded 90°-specimen.

As for the investigations in the injection molding direction, the curves of the accumulated acoustic events are also compared here. Figure 6.12 shows non-preloaded specimens (a) and the preloaded ones sorted by the maximum stress during preloading (b). The curves of the accumulated acoustic events show the same behavior as those of the 0°-specimens. Below a certain maximum stress during the preloading, the tendency of the curves changes and approaches those of the non-preloaded curves. For the 90°-specimens this is first observed for the specimen preloaded with a maximum stress of 35% of R_{UTS}.

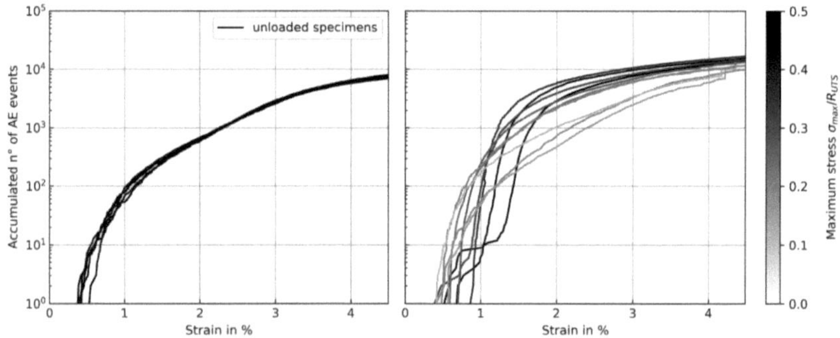

(a) (b)

Figure 6.12: Accumulated number of acoustic events of 90°-specimens: (a) non-preloaded; (b) preloaded sorted by maximum relative stress level during preloading.

Figure 6.13 shows the accumulated AE events for the non-preloaded specimens and the specimens preloaded with the four lowest maximum stresses to determine the preloading condition where the AE activity of the preloaded specimens equals that of the non-preloaded specimens. For a maximum stress during preloading equal to or below 20% of R_{UTS} the curves of accumulated acoustic events run as those of the non-preloaded specimens. This suggests that the applied cyclic load did not cause any measurable damage. Accordingly, the AE results of the residual strength tests indicate that for the investigated material PA6.6GF40 transverse to the injection molding direction, a fatigue limit exists at the earliest for maximum stresses less than or equal to 20% of R_{UTS} or, with respect to the number of load cycles, for more than $2 \cdot 10^{10}$ load cycles. This is consistent with the results of the stiffness degradation studies in Section 5.4. Likewise, stiffness degradation was measured here for a maximum stress greater than 20% of R_{UTS}, indicating that no endurance limit exists above this limit. The stress-strain curves and the acoustic activity of the individual tests on the 90°-specimens are shown in Appendix A.5.

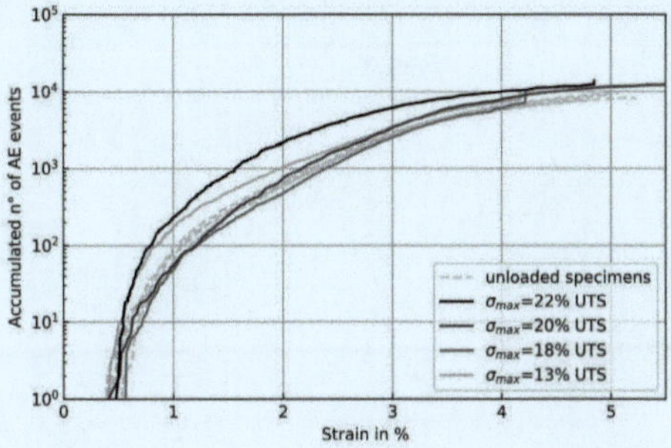

Figure 6.13: Accumulated number of acoustic events of 90°-specimens, non-preloaded and preloaded with σ_{max} = 22%, 20%, 18%, 13% RUTS.

7 Digital twin simulation

In the following, the modeling of a digital twin with the software GeoDict® [9] is presented. The objective of the modeling is to investigate the micromechanical mechanisms underlying the critical stress levels, which are experimentally determined in Chapters 5 and 6. If this proves successful, the results can be transferred to other microstructures, such as other fiber contents or other fiber orientation distributions, using the digital twin.

For this purpose, the procedure for analyzing the microstructure is explained in a first step in Chapter 7.1. Subsequently, in Chapter 7.2 the generation of the digital twin is described. This model is used to simulate the quasi-static tensile tests in (0°) and transverse to (90°) the direction of injection molding, which is presented in Chapter 7.3. Finally, the simulations of the tensile tests with the generated model are discussed in Chapter 7.3.4.

7.1 Analysis of the microstructure

The fiber content is determined by ashing, in a first step. Subsequently, two samples with the dimensions of the gauge section (3 mm x 8 mm) are scanned in an X-ray (520 Versa, Zeiss) for the analysis of the microstructure. The resolution is set to 2 µm, in order to be able to map the fiber diameter with several voxels. The scan images are then imported in the software GeoDict®, in order to estimate the fiber orientation, the fiber diameter and the fiber length distribution.

7.1.1 Fiber content

In order to verify the fiber content of 40 wt.-% specified by the manufacturer, three samples are taken from each of the three injection-molded plates at different positions of the plate to capture any variation of the fiber content that may occur (see Figure 7.1). The samples are then dried until the mass changes by less than 0.1% within one day. Subsequently, the samples are heated in an oven to 550°C to burn the PA6.6 and thus release the fibers from the matrix.

Figure 7.2 shows the data and the mean value of the fiber content at the three positions. The fiber content decreases with increasing distance from the sprue point, from 39.5 wt.-% at the top to 38.9 wt.-% at the bottom. The overall mean value of the fiber content was estimated to 39.2 wt.-%.

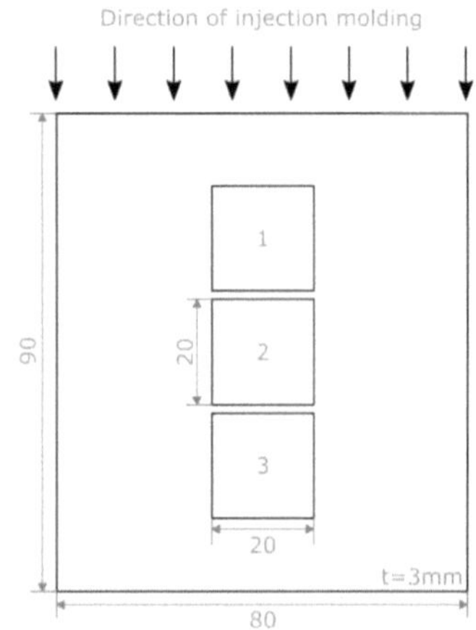

Figure 7.1: Sample position on plate for verification of the fiber content.

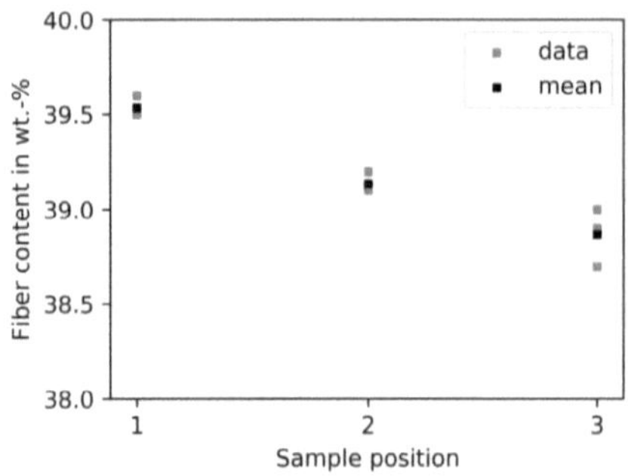

Figure 7.2: Test data and mean value of fiber content for the three sample positions.

7.1.2 Fiber orientation

The X-ray scans of two samples are imported into GeoDict® for further analysis of the microstructure. Due to the high density difference of glass fibers and the PA6.6 matrix, the two materials have different gray values and can be distinguished from each other on this basis. The exact threshold value for differentiating between the two components is manually adjusted so that the fiber content matches the determined average value. The software offers numerous image filters to sharpen the contrast between the two phases, which adjust the brightness when the gray value flicker occurs in one or more directions, and among which the so-called *gradient brightness correction* is successfully applied in all three directions and to all gray value ranges.

For the estimation of the fiber orientation, in a first step, the fibers are identified using image processing techniques based on gray values. Afterwards, the fiber orientations are computed for each voxel using the so-called *Star Length Distribution* algorithm. The average fiber orientation tensor is determined as follows, with the x-axis pointing in the injection molding direction:

$$a = \begin{bmatrix} 0.54 & 0.00 & 0.00 \\ - & 0.37 & 0.00 \\ - & - & 0.09 \end{bmatrix} \quad (7.1)$$

The fiber orientation tensor shows that the fibers are mostly oriented in the x-y plane, and that hardly any fibers are oriented in the thickness direction. It also illustrates that more fibers are oriented in the direction of injection molding, than transverse to the direction of injection molding.

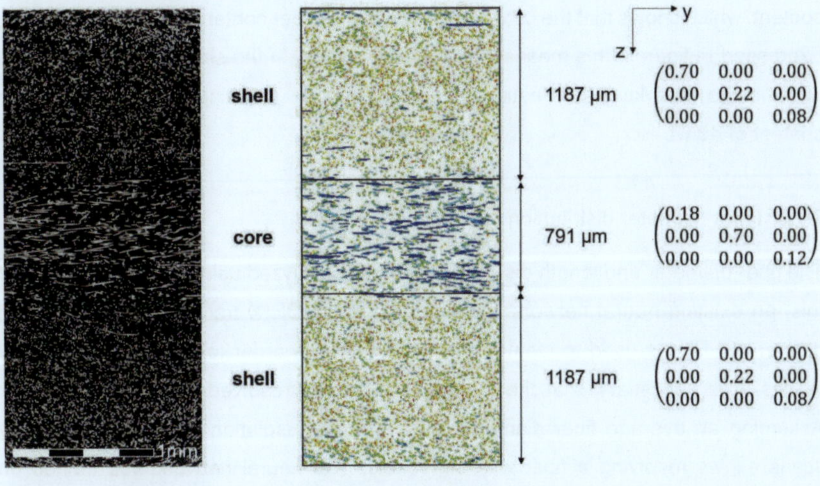

Figure 7.3: (a) X-ray scan with a resolution of 2.0 µm; (b) identified fibers colored by fiber orientation; (c) estimated fiber orientation tensors. Direction of injection molding is along the x-axis.

In addition to the mean fiber orientation tensor, the fiber orientation distribution is investigated further. The x-ray scan shows the typical through the thickness core-shell morphology of injection-molded parts (Figure 7.3 (a)). The scan also shows two skin layers, which for simplicity will be considered part of the shell layers for further considerations. The thickness of the shell and the core layers is determined by splitting the scan into eight bins across the thickness and analyzing the main fiber orientation of these bins. The orientation tensors of these bins are similar for the outer three bins and the two middle bins (Figure 7.3 (b)). With this approach, the fiber orientation tensors of the shell and core layers result to:

$$a_{shell} = \begin{bmatrix} 0.70 & 0.00 & 0.00 \\ - & 0.22 & 0.00 \\ - & - & 0.08 \end{bmatrix}, a_{core} = \begin{bmatrix} 0.18 & 0.00 & 0.00 \\ - & 0.70 & 0.00 \\ - & - & 0.12 \end{bmatrix} \quad 7.2$$

While the majority of the fibers in the shell layers are oriented in the injection molding direction, the main orientation in the core layer is transverse to the injection molding direction. Overall, however, the proportion of fibers oriented in the direction of injection molding predominates, since the shell layers are thicker than the core layer.

The fiber orientation analysis in GeoDict® also includes the analysis of the fiber content, which shows that the core layer has a higher fiber content than the shell layers. Expressed in figures, this means that the fiber content in the shell layers is 37 wt.-%, while in the core layer it is estimated to be 49 wt.-% resulting in an average fiber content of 40 wt.-%.

7.1.3 Fiber diameter distribution

The fiber diameter and length distribution are both analyzed using AI-approaches. For this, an existing neural network implemented in GeoDict® trained on µCT scans of different SFRTs is used to identify individual fibers. In order to achieve good quality results, it is necessary, that the individual fibers are resolved with multiple voxels. Assuming an average fiber diameter of 12 µm, the resolution of 2 µm of the X-ray scans allows resolving a fiber with six voxels. The neural network was trained on circular fibers and therefore works only for this fiber shape.

In each X-ray scan, the software counts about 10^5 fibers and evaluates the dimensions of the individual fibers. Figure 7.4 (a) shows the probability distribution of the fiber diameters. It is visible that the distribution is bell-shaped and nearly symmetrical to two peaks at 10.5 and 11.0 µm. Additionally, the Gaussian distribution can approximate the distribution of fiber diameters well with a mean value of μ = 10.9 µm and the standard deviation of $\sigma^2 = 1.0$ µm.

The fiber diameter distributions can also be computed by image processing techniques. Matrix or fiber material is assigned to each voxel based on its gray values for this. Subsequently spheres are fitted to the fiber voxels to iteratively estimate the fiber diameter. The results are shown in Figure 7.4 (b) and indicate that this method yields both a higher mean fiber diameter and a larger standard deviation than the evaluation with AI. Using image-processing techniques, the average diameter is estimated to be 11.8 µm with a standard deviation of 2.7 µm. Since the fiber diameter distribution estimated with AI correlates well with literature data, it can be assumed that this technique leads to more accurate results [99,100].

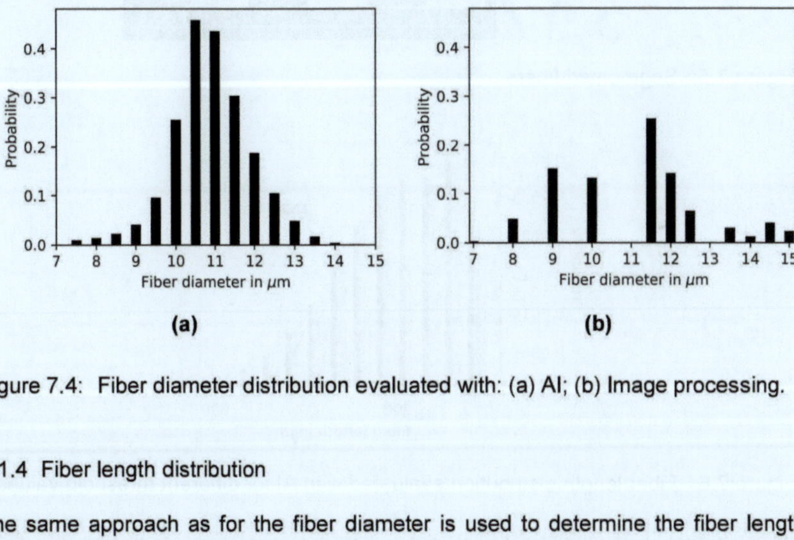

Figure 7.4: Fiber diameter distribution evaluated with: (a) AI; (b) Image processing.

7.1.4 Fiber length distribution

The same approach as for the fiber diameter is used to determine the fiber length distribution. Before evaluating the fiber length, it is necessary to check that the algorithm does not over-segment the fibers. This check is performed purely visually by displaying the fibers in different colors. Figure 7.5 shows the segmented fibers, and illustrates that some longer fibers are by mistake split into multiple fibers. It can therefore be assumed that the average fiber length is underestimated.

A threshold that decides which material is assigned to a voxel can control the over-segmentation. Figure 7.6 shows the influence of this threshold value on the evaluated fiber length distribution. It becomes clear that the proportion of short fibers decreases as the threshold value decreases, which also indicates that the algorithm tends to identify too many fibers, resulting in an over-estimated standard deviation.

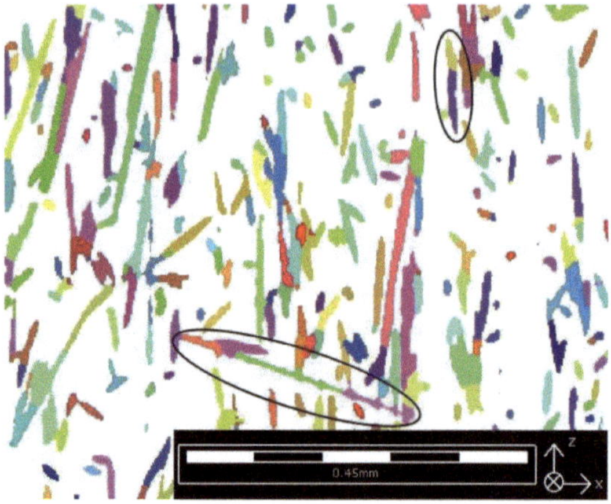

Figure 7.5: Segmented fibers.

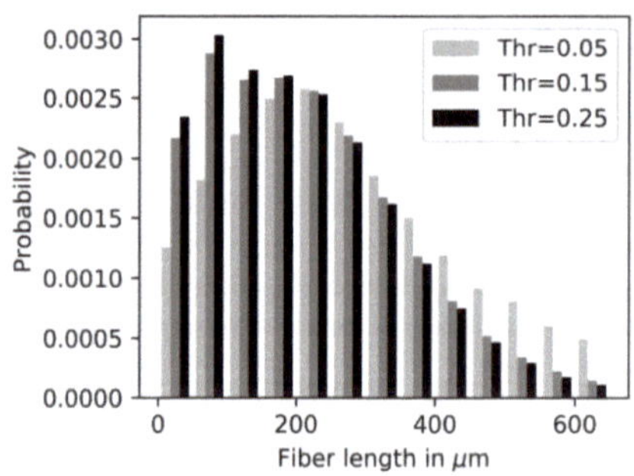

Figure 7.6: Fiber length distribution evaluated with AI for different threshold values.

Figure 7.7 (a) shows the fiber length distribution determined with AI (threshold = 0.05). The mean fiber length is estimated to be 333 μm, with a standard deviation of 264 μm. In order to rule out the possibility that this result is due to the previously explained over-segmentation, the fiber length distribution is also determined using microscopic images. In the images, 1300 fibers are counted and measured manually. The histogram of this measurement is shown in Figure 7.7 (b), together with a Gaussian distribution with a mean of $\mu = 227.0$ μm and the standard deviation of $\sigma^2 = 115$ μm.

When comparing the two determined fiber length distributions, it is noticeable that the mean and standard deviation differ significantly. However, for both methods, fibers with a length between 150 and 250 µm account for the largest proportion. The higher standard deviation of the determination with AI can be explained by the greater proportion of short fibers, which is related to the over-segmentation. Table 7.1 lists the mean value and the standard deviation of the fiber length determined with the different methods.

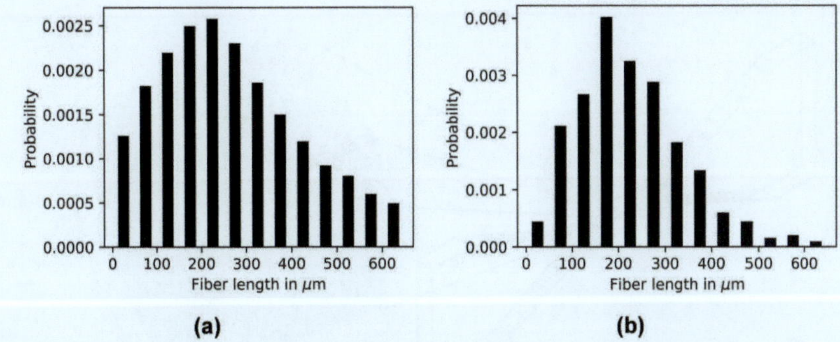

Figure 7.7: Fiber length distribution evaluated with: (a) AI; (b) Microscopic images.

Table 7.1 Mean value and standard deviation of fiber length distribution

Method	Threshold	Mean fiber length in µm	Standard deviation of fiber length in µm
Microscopic measurement	-	227	115
AI	0.05	333	264
AI	0.15	220	153
AI	0.25	209	144

7.2 Generation of the digital twin

As requirements for the generated microstructure, the volume has to be as small as possible to minimize computation time and at the same time as large as necessary to represent all relevant parameters of the real microstructure. Since the parameters affecting the prediction quality are different in the three spatial directions, individual model dimensions are also chosen in the x, y, and z directions. While the fiber length distribution has a major influence on the model quality in and transverse to the injection molding direction, its influence in the thickness direction can be neglected. In order to incorporate long fibers into the digital twin, the edge length in the plane (x- and y-direction) is 400 voxels, which corresponds to 800 µm. In the thickness (z-) direction, on the other hand, the fiber orientation distribution plays a major role, which is reflected in the core-shell morphology. For this purpose, the shell area is mapped with 119 voxels and the core area with 79 voxels, which corresponds to a total of 20% of the thickness. Figure 7.8 shows the dimensions and structure of the digital twin.

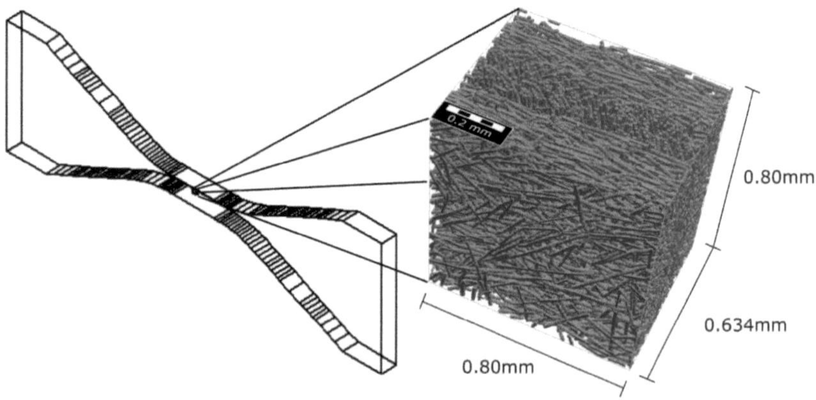

Figure 7.8: Dimensions of digital twin.

Once the dimensions of each layer are determined, the scanned structure is examined region by region to find areas of representative microstructure. The first requirement for these representative areas is a matching fiber content with the average fiber content in the respective plies. A second focus is on the fiber orientation tensors determined for the core and shell plies (see Chapter 7.1.2).

After identifying a representative bin for the shell and core layer, the individual fibers are identified with AI. The resulting geometric properties of the identified fibers, such as the dimensions and the position, are written to an ASCII-file. This allows the properties of the individual fibers to be processed separately and fiber overlaps to be detected. The last step is to merge the processed structure of the shell bin on each side of the processed structure of the core bin (see Figure 7.8). In consistency with the real specimens, the fibers in the shell layer are oriented mainly in the direction of injection molding, while they are oriented mostly transverse to the direction of injection molding in the core layer. Figure 7.9 illustrates the method developed in this work to generate the digital twin microstructure using GeoDict®.

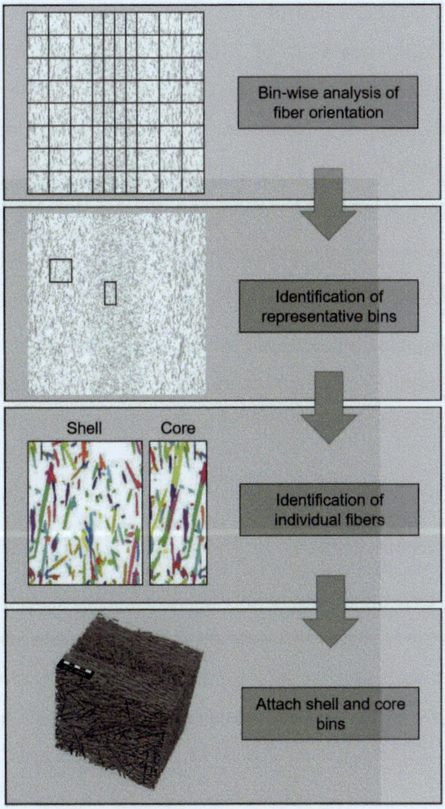

Figure 7.9: Method for generating the digital twin.

In the following, the microstructural properties of the digital twin are compared with those of the scanned specimen. The fiber content of the digital twin is 39.9% by weight and thus agrees with the average fiber content of the test specimens. When comparing the average fiber orientation tensor of the scan (eq. 7.1) with that of the digital twin (eq. 7.3), it is noticeable that the digital twin has a slightly stronger orientation of the fibers in the flow direction.

$$a = \begin{bmatrix} 0.58 & -0.01 & -0.02 \\ - & 0.35 & 0.02 \\ - & - & 0.06 \end{bmatrix} \qquad 7.3$$

Using an AI approach, the mean fiber diameter of the digital twin is determined to 11.3 µm with a standard deviation of 1.2 µm which also matches the fiber diameter distribution of the specimens. The fiber length distribution of the digital twin is determined with a threshold value of 0.05. This determines the mean fiber length of 193 µm with a standard deviation of 210 µm. In the fiber length distribution, the influence of the smaller dimensions of the digital twin compared to the real specimen is most apparent. At this point, however, modeling accuracy and computation time must be balanced. Nevertheless, the digital twin reproduces the real structure overall with sufficient accuracy to assess the evolution of local stresses and strains.

7.3 Simulation of the quasi-static tests

The experimental results are presented in this chapter, followed by the material laws used for the simulation. Finally, simulation results are presented and discussed to address the critical stress levels identified in Chapters 5 and 6 at the microstructural level.

7.3.1 Experimental tensile tests

The tests are conducted according to DIN EN ISO 527-1 [97] with a testing velocity of 1.0 mm/min. The strain is measured with the DIC-system Aramis. Two series of tests with five specimens each are carried out, one with specimens in and the other with specimens transverse to the injection molding direction. See Chapter 6.1.2 for more information.

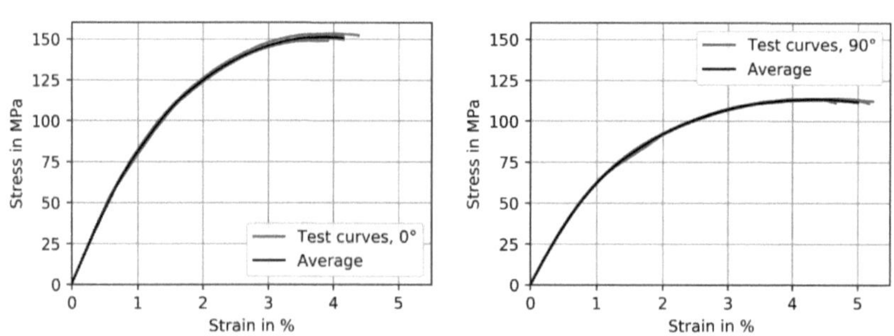

(a) (b)

Figure 7.10: Stress-strain curves: (a) 0°-specimens; (b) 90°-specimens.

The experimental stress-strain curves with the calculated average of the test curves are presented in Figure 7.10. Due to the higher fiber orientation in the direction of injection molding, the 0°-specimens have a greater stiffness and strength than the 90°-specimens. Both test curves show an elastic-plastic behavior, but the 90°-specimens yield a greater failure strain than the 0°-specimens.

7.3.2 Material laws

The material models chosen for the glass fibers and the PA6.6 matrix are explained below. It is pointed out at this point that the interface between fibers and matrix was not considered within the scope of this work.

Glass fibers

An elastic material law is used with the parameters provided by the supplier for the glass fibers (Nippon Electric Glass Co., Ltd). Since the fibers show a brittle failure behavior, a failure model with a failure variable that can be either zero or one is chosen. As soon as the tensile failure stress in a voxel is reached, the failure variable is set to one and the stiffness of the voxel in question is abruptly reduced to 1% of the initial stiffness.

Table 7.2 Material properties of glass fibers

Parameter	Value
Young's modulus	73 GPa
Poisson ratio	0.18
Failure Stress	2.6 GPa

Polyamide 6.6

Since the PA6.6 shows a pronounced plasticity, it is modeled with an elastic-plastic material law. The plastic region of the material is described by exponential hardening with the following correlation between the strain and the stress:

$$\sigma(\varepsilon_{plastic}) = Y + \alpha \cdot (1 - e^{-\beta \varepsilon_{plastic}}) \qquad 7.4$$

Where Y stands for the Yield stress, α and β are the first and the second hardening parameter and $\varepsilon_{plastic}$ is the plastic strain. The parameters of the law Y, α and β are used to fit the digital twin simulation to the experimental test data. Figure 7.11 shows a schematic presentation of the elastic-plastic material law with exponential hardening and Table 7.3 shows the parameters of the elastic-plastic material law of PA6.6 used for the digital twin simulation.

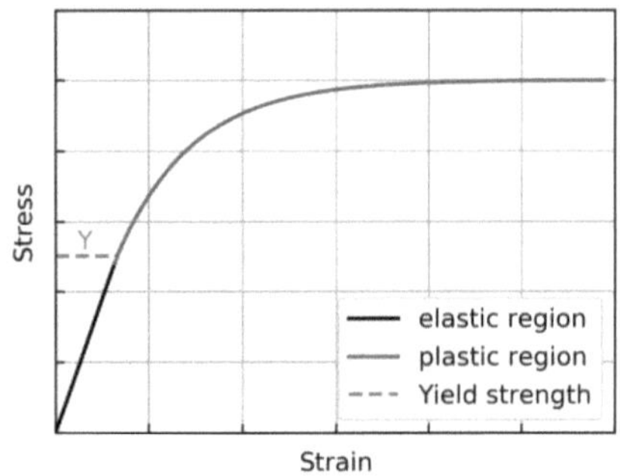

Figure 7.11: Schematic elastic-plastic material law with exponential hardening.

Table 7.3 Elastic-plastic material law of PA6.6

Parameter	Value
Young's modulus	2.5 GPa
Poisson ratio	0.35
Yield stress	0.03 GPa
First hardening parameter	0.075
Second hardening parameter	15

7.3.3 Stiffness

The symmetry of the core-shell morphology is used by running the calculation only on half of the digital twin structure to save computational power. In a first step, the quality of the digital twin is assessed by comparing the simulated stiffness with the Young's modulus measured in the quasi-static tensile tests. The Young's modulus E_t is determined for the linear elastic region using the formular:

$$E_t = \frac{\sigma_2 - \sigma_1}{\varepsilon_2 - \varepsilon_1} \qquad 7.5$$

DIN EN ISO 527-1 [97] defines the linear elastic region between a strain of 0.05 % and 0.25 %. Table 7.4 shows the mean values of the Young's modulus determined in the quasi-static tensile tests and with the digital twin simulation illustrating that the digital twin well simulates the stiffness of the SFRT, especially when considering the more pronounced fiber alignment of the digital twin.

Table 7.4 Tensile Young's modulus of PA6.6GF40

Θ to direction of injection molding in °	Experimental E_t in MPa	Digital twin E_t in MPa
0	9318	9775
90	7468	7051

7.3.4 Simulated tensile tests

The simulated stress-strain curves are shown in Figure 7.12 together with the experimental stress-strain curves. While the digital twin simulation transverse to the injection molding direction reproduces the experimental test curves very well, a bigger offset from the test curves can be observed for the simulation in the direction of injection molding.

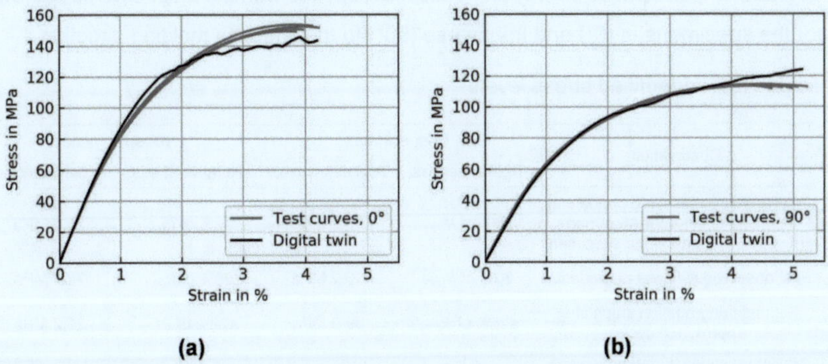

Figure 7.12: Stress-strain curves of experimental testing and digital twin simulations: (a) 0°-specimens; (b) 90°-specimens.

At approx. 1.5% strain, the simulated 0° curve flattens and becomes jagged in the following, which is not observed in the simulation of the 90°digital twin. This can be explained by the more pronounced fiber failure in the 0° digital twin and the size of the simulated twin. As soon as a single fiber fails, this is reflected in the stress-strain curve of the digital twin, since only a few fibers are simulated, and the influence of a single fiber failure is therefore more pronounced. However, since this work focuses on low stresses in the fatigue range, the digital twin simulation is accurate enough for conclusions on the mechanisms prevailing in this stress range. The local stress state in the fibers and the matrix is discussed in the following.

7.4 Discussion of the simulation

The following section examines the results of the digital twin at the stress levels identified in Sections 5.4 and 6.4 as proven characteristic of fatigue behavior. In Section 5.4, the evaluation of data obtained during fatigue tests showed that below a certain stress level no stiffness degradation is measurable. A second level was identified where stiffness degradation ceased after the first approximately 10^3 cycles. Likewise, an interval was determined, in which the stiffness degradation and the dissipative energy depend linearly on the maximum stress. The examination of AE during quasi-static tests in Section 6.4 revealed a maximum stress level below which the acoustic response of cyclically preloaded specimens is comparable to the acoustic response of non-preloaded specimens. Table 7.5 lists the observations on stiffness degradation, dissipative energy and acoustic response with the respective stress levels for the specimens in (0°) and transverse (90°) to the injection molding direction.

Table 7.5 Identified stress levels

Observation	0°-specimens		90°-specimens	
	Relative stress	Absolute stress	Relative stress	Absolute stress
Maximum stress during preloading resulting in different AE of non-preloaded and preloaded specimens	20% of R_{UTS}	30.2 MPa	22% of R_{UTS}	25.2 MPa
First observed stiffness degradation	20% of R_{UTS}	30.2 MPa	20% of R_{UTS}	22.9 MPa
Stiffness degradation stops after approx. 10^3 cycles	≤26% of R_{UTS}	39.9 MPa	≤22% of R_{UTS}	25.2 MPa
Maximum stress during preloading resulting in AE approaching the AE of the non-preloaded specimens	≤26% of R_{UTS}	39.9 MPa	≤35% of R_{UTS}	40.0 MPa
Stiffness degradation and dissipative energy depend linearly on the maximum stress level	48%-57% of R_{UTS}	73.2-86.5 MPa	45%-57% of R_{UTS}	51.4-65.4 MPa

With the chosen material laws in the realized digital twin simulation, the evolution of two parameters can be investigated: the plasticization of the matrix and the fiber failure. In a first step, the evolution of the plasticization of the matrix material with increasing external load is examined. For this purpose, the matrix volume that starts to plasticize in each calculation step is evaluated. Plotting this plasticized volume versus global strain shows that the plasticization rate resembles a cumulative Gaussian distribution that is consistent for simulations both in and transverse to the injection molding direction (Figure 7.13). The probability density function with the parameters μ and σ as the mean and the standard deviation of the distribution is expressed by:

$$f(x) = \frac{1}{\sigma\sqrt{2\pi}} e^{-\frac{1}{2}\left(\frac{x-\mu}{\sigma}\right)^2} \tag{7.6}$$

The cumulative plasticized matrix volume is analyzed using a least-square minimization implemented in *python's SciPy* library. Table 7.6 lists the determined mean strains and standard deviations as well as the corresponding stresses. With a coefficient of determination of $R^2 = 0.983$ (0.979) for the digital twin in (transverse to) the direction of injection molding, the Gaussian distribution closely fits the data. This is also evident from Figure 7.14, which furthermore shows that especially the first half up to reaching the maximum is accurately modeled by the Gaussian distribution.

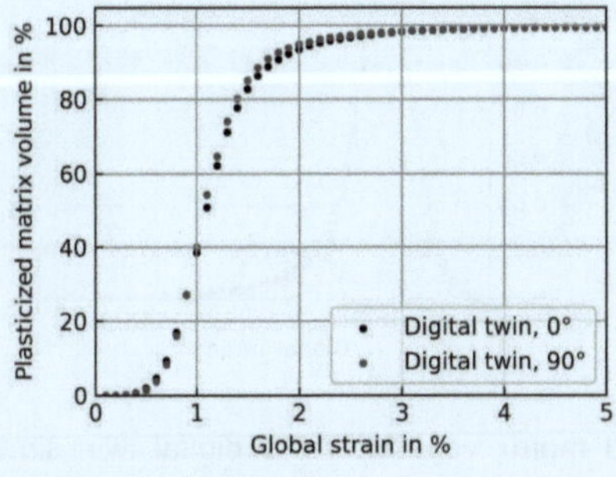

Figure 7.13: Cumulative plasticized matrix volume in the digital twin simulations.

Table 7.6 Gaussian distribution of plasticized matrix material over global strain

Θ to direction of injection molding in °	μ in %	σ in %
0	1.100	0.306
90	1.087	0.281

Besides being symmetric about the mean μ that thus indicates the global strain at which half of the matrix volume is plasticized, the Gaussian distribution has the property that its two inflection points are located one standard deviation σ away from the mean μ at $x = \mu \pm \sigma$.

Table 7.7 lists global stresses at the means and inflection points of the estimated Gaussian distributions. Comparing these stresses with the listed characteristic stress levels in Table 7.5, it is noticeable that the means and first inflection points coincide with the stress values determined from analyzing stiffness degradation and dissipative energy, which were found to delimit the HCF range. This observation using the digital twin is interpreted to mean that in the HCF range, the plastic material behavior of the matrix plays an important role, while it can be neglected in the VHCF. This is also in agreement with the observations and conclusions of CARVELLI, JAIN AND LOMOV [94], who observe mainly fiber-matrix delamination in the region of VHCF, thus identifying the failure of the fiber-matrix interface as the governing damage mechanism in VHCF. This also explains why no correlation to the plasticizing behavior of the matrix is apparent for the other characteristic stress levels in Table 7.5.

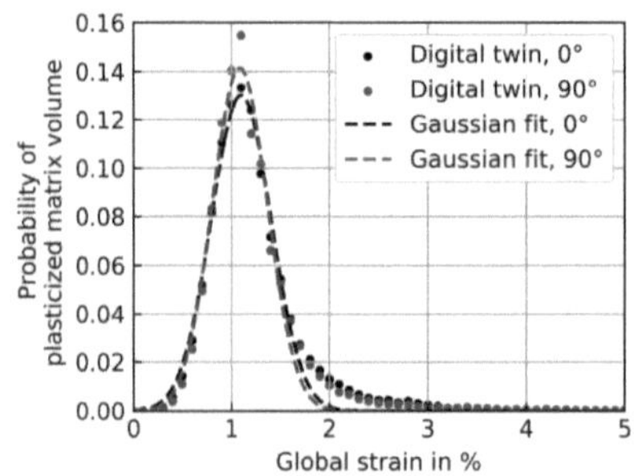

Figure 7.14: Plasticized matrix volume in the digital twin simulations and Gaussian distribution.

Table 7.7 Corresponding global stress at mean and inflection points of estimated Gaussian distributions of plasticized matrix material

Θ to direction of injection molding in °	μ in MPa	1st inflection point in MPa	2nd inflection point in MPa
0	96.5	74.1	114.3
90	68.6	54.8	80.4

In a second step, the evolution of fiber failure in the digital twin simulations is investigated. Figure 7.15 shows the cumulated volume in which the simulations predict fiber failure. The graph visualizes, that the fiber failure starts at higher global strains than the plasticization of the matrix. Unlike the matrix plasticization rate, the fiber breakage rate is almost constant. Therefore, a linear approximation is used to determine the global strain at which fiber failure starts. Based on the corresponding global stress values at the onset of fiber failure listed in Table 7.8, it can be concluded, that fiber failure only affects the damage behavior in LCF.

The observations and conclusions on the failure behavior of SFRT explained above illustrate that the accuracy of the model is insufficient to adequately investigate the domain of VHCF. For this purpose, it is necessary to consider the fiber-matrix interface in the digital twin simulation. The modeling of the fiber-matrix interface, however, results in two major challenges: first, the resolution has to be chosen fine enough to represent the interface with at least one voxel. This in turn leads to either large models with long computation times or a small volume that can be modeled. Second, the availability of material data. While the interfacial properties for continuous fiber-reinforced materials can only be determined with high testing effort and large result scatter, a characterization of the fiber-matrix interface of short fiber reinforced materials has hardly been examined so far. This is primarily due to the fact that fibers and matrix are already mixed in the injection molding granulate and are rarely available separately. Also, the handling of the short fibers is challenging, for example in fiber pull-out tests leading to even more pronounced scatter than in the characterization of the interfacial properties of continuous fiber-reinforced polymers.

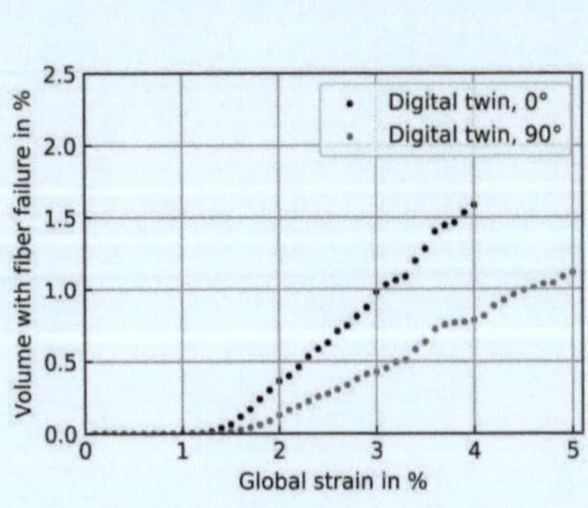

Figure 7.15: Cumulative fiber failure volume in the digital twin simulations.

Table 7.8 Global strain and corresponding global stress at start of fiber failure

Θ to direction of injection molding in °	ε_0 in %	σ_0 in MPa
0	1.44	115.7
90	1.65	89.5

8 Conclusion

The pronounced anisotropic behavior of short fiber reinforced thermoplastics and the various factors influencing the mechanical properties, such as fiber length distribution and environmental conditions, lead to a high effort for comprehensive fatigue characterization. Therefore, the aim of the presented work is to explore experimental methods that allow obtaining information on the fatigue behavior of SFRT in a time-efficient manner. On the one hand, this enables shorter development times and, on the other hand, results in an improvement of the design process, which is often based on static rather than fatigue investigations for reasons of time and cost.

This work pursues the approach of using the potential of modern computer technology for processing and analyzing large amounts of data as well as for virtual testing. All investigated methods have the common feature that they generate big data sets that do not allow for manual evaluation but instead rely on computer-aided analysis methods. In this context, the work examines three different approaches: the analysis of data acquired during fatigue tests, acoustic emission testing, and virtual testing by simulating a digital twin.

For the experimental investigations, a uniform specimen geometry is optimized to meet all the requirements of the different test methods. The aim of optimization is to avoid stress peaks caused by the specimen geometry in order to provoke a failure due to micromechanical influences. This goal is opposed by the fact that high-resolution X-ray microscopic scans for analyzing the microstructure of specimens can only image a limited volume on the one hand, and on the other hand, performing AE inspections requires a complete application of AE sensors. These conflicting requirements on the specimen width lead to a significant widening of the cross-section from the measurement length to the cross-section where the sensors are applied. To optimize a smooth transition from the narrow to the wide cross section, the software LS-OPT® from DYNAmore is used resulting in a Bézier-shaped widening of the cross-section. The test specimens are milled from injection molded sheets of polyamide 66 with 40% by weight short glass fiber reinforcement. In order to detect an effect of the fiber orientation, specimens are produced in and transverse to the injection molding direction.

As a basis for further investigations, tensile-tensile cyclic tests are carried out with the specimens in 0° and 90° to the injection molding direction in order to create the corresponding S-N diagrams. Specimens that do not fail before 10^7 load cycles are referred to as runouts and not considered for the estimation of the S-N curve. Subsequently, further tensile-tensile fatigue tests are carried out in the HCF and the VHCF range with machine data recorded. The advantage of using machine data is that it can be acquired without the use of additional test equipment. The recorded machine data is used to determine the stiffness degradation and dissipative energy during cyclic loading, as these two parameters are directly linked to the damage behavior in the material.

It is shown that the stiffness degradation over the lifetime can be expressed by two logarithmic curves. The slopes of these curves illustrate that the stiffness decreases faster in the first cycles up to approximately 10^3 cycles. Subsequently, it is investigated whether and up to which number of load cycles the fatigue tests can be shortened so that the same parameters for the stiffness degradation curves can still be determined. These studies show that when tested up to 10^5 load cycles, a uniform condition occurs that allows extrapolation of stiffness degradation. As a result, the test times can be significantly reduced compared to the loading to failure of the specimens.

In a next step, the slope of the first and the second part of the stiffness degradation curves are analyzed with respect to the applied maximum stress. The investigations reveal that the stiffness degradation can be divided into three ranges depending on the slope parameters. In each of these ranges a linear relationship can be established between the applied maximum stress and the stiffness degradation rate. In this way, two characteristic stress levels are identified. Below the first characteristic stress level, there is only a minor stiffness degradation detected, with most of it occurring in the first 10^3 load cycles. Between the first and second characteristic stress level stiffness degradation increases significantly. For stresses higher than the second identified stress level, a jump in the rate of stiffness degradation is observed, particularly in the second part of fatigue loading after 10^3 load cycles. For the characteristic stress levels determined in this way, the corresponding load cycles to failure are estimated using the S-N diagrams prepared previously. For both slope parameters and for both orientations, these limits are consistent with the usual limits of the HCF range between 10^4 and 10^6 load cycles. From these findings on stiffness degradation, it can be concluded that different damage mechanisms prevail in the LCF, HCF and VHCF regions. While in HCF damage propagates constantly, in the VHCF region damage occurs at individual stress peaks, e.g. at fiber ends or between adjacent fibers, but the energy is not sufficient for this damage to grow further.

If no stiffness degradation is measurable, this can serve as an indicator of the existence of an endurance strength below which no fatigue failure occurs. For the specimens in the injection molding direction no stiffness degradation is measured below a stress with a corresponding fatigue life of 10^{11} load cycles. For the specimens transverse to the injection molding direction this occurs for stresses for which the fatigue life is extrapolated to 10^{10} load cycles. At a test frequency of 10 Hz a fatigue life of 10^{11} load cycles corresponds to a test duration of approx. 300 years, which illustrates that experimental validation with conventional test methods is not possible.

Additionally to the stiffness degradation, hysteresis behavior is studied. Using the hysteresis data, the dissipative energy is determined, which shows a large scatter compared to the stiffness data. Therefore, no functional relationship between the dissipative energy and the number of load cycles can be derived, but only an average dissipative energy can be determined. However, when visualizing this mean dissipative energy versus the applied maximum stress, the same trend as for the stiffness degradation rate becomes apparent. The mean dissipative energy can also be used to identify the characteristic stress levels that delimit the HCF from the LCF and VHCF regions. In contrast to stiffness degradation, dissipative energy is measurable even at the lowest stress levels, which on the one hand supports the previous assumptions about damage behavior in VHCF and on the other hand contradicts the existence of an endurance limit of SFRT.

The second part of the experimental investigations consists of quasi-static tensile tests with acoustic emission analysis. For this, two acoustic sensors are used during mechanical testing to detect acoustic signals generated by damage events such as matrix cracking or fiber breakage. Since processing the large amounts of data generated in AE is challenging, AE has so far only been applied in quasi-static tests and short-term fatigue tests. In order to study the damage behavior in the VHCF range, this work investigates the acoustic response in residual strength tests. The specimens are preloaded with different maximum stresses and different number of load cycles before the AE is measured in residual strength tests. The idea behind this is that damage that has already occurred during preloading can no longer be measured during residual strength testing. AE of these residual strength tests is then compared to the AE during quasi-static tests on non-preloaded specimens. A matching AE response of non-preloaded and preloaded specimens may again be indicative of an endurance limit.

Since the amount of acoustic data is very large, several filtering steps are required before further analysis. For this purpose, a custom *python* script reduces the data until only those acoustic events remain that occurred in the measurement cross section. Subsequently, the progression of the cumulative acoustic events over the applied global strain is analyzed. This shows that for the specimens in the direction of injection molding the acoustic response of the non- and the preloaded specimens agree for a maximum stress less than or equal to 18%. Extrapolating the determined S-N curve, the corresponding fatigue life for this maximum stress is 10^{11}. Consequently, no endurance limit exists below 10^{11} load cycles, which is consistent with the findings obtained from analyzing the stiffness degradation data. The results of the AE tests on the specimens transverse to the injection molding direction also lead to the same upper limit for an endurance limit as the stiffness degradation tests at 10^{10} load cycles.

Additionally to the residual strength tests with different maximum stresses during preloading, the influence on the acoustic response for a preloading with different number of load cycles while maintaining the same maximum stress is investigated. The results are consistent with the assumptions on damage initiation and propagation derived from the stiffness degradation analysis. For a preloading with 10^3 and 10^4 load cycles, a number of small strain acoustic events is measured during the residual strength tests. In comparison, in the quasi-static tests on non-preloaded specimens the first AE events occur only at significantly higher strains. These small strain events are interpreted as being caused by damage that has already been initiated by the fatigue preloading and therefore requires only a small amount of deformation energy to propagate further. For a preloading with 10^5, 10^6, and 10^7 load cycles, the acoustic events start after passing the strain applied during the preloading. The investigation also shows that the acoustic response in the residual strength test of the specimen preloaded with 10^2 load cycles is equal to the AE of the non-preloaded specimens apart from single AE events at low strains. Therefore, it can be concluded that fatigue loading with a number of load cycles of less than 10^3 cycles does only cause sporadic damage. For a preloading between 10^3 and 10^5 load cycles, the group of acoustic events with low amplitude at a small strain indicates that micro-cracks develop between 10^3 and 10^5 load cycles, which only require low deformation energy to grow further in the residual strength tests. For a preloading above 10^5 load cycles, no damage events occur at low strains indicating that a steady damage state is reached after a preloading with 10^5 load cycles.

To investigate the micromechanical mechanisms underlying the experimentally determined characteristic stress levels, a digital twin is simulated with the commercial software *GeoDict®*. For this purpose, X-ray microscopy scans of the test specimens are used, which are imported into and analyzed by the software. The fiber content, orientation, diameter, and length distribution are evaluated, in order to create a twin with a representative microstructure. This is achieved by joining part of the shell to part of the core layer, focusing on an overall realistic fiber orientation.

The quasi-static tensile tests in and transverse to the injection molding direction are modeled assuming that observations on damage behavior under quasi-static loading allow conclusions on the fatigue behavior under cyclic loading. While an elastic-plastic material law is chosen for the matrix material, the fibers are modeled with an elastic material law combined with a failure model. The analysis of plasticization rate of the matrix material shows that a Gaussian distribution, whose mean and standard deviation allow an estimation of the stresses that limit the HCF range, can approximate it.

The simulation also shows that fiber failure starts at higher global strains and stresses than the plasticization of the matrix. Unlike the matrix plasticization rate, the fiber breakage rate is approximately constant. The global stress values at the start of fiber failure suggest that fiber failure only affects the damage behavior in LCF.

However, the chosen simulation does not show any specific micromechanical damage mechanisms at low stresses which correspond to the VHCF range. It is assumed that this is due to the fact, that the fiber-matrix interface dominates the damage behavior in this fatigue range. Furthermore, when evaluating the model in the other fatigue domains such as LCF and HCF, it must be taken into account that the present model does not represent the behavior of the fiber-matrix interface and its possible influence on the damage process.

In summary, the chosen analysis method of stiffness degradation and dissipative energy allows to estimate the HCF range of SFRT. Furthermore, it is shown that a fatigue loading up to 10^5 load cycles is sufficient for this purpose, allowing a time-efficient estimation of the fatigue properties to be achieved. These results are verified with a digital twin simulation, which simultaneously provide insight into the underlying micromechanical mechanisms. Stiffness degradation analysis and residual strength tests with AE indicate that an endurance limit for SFRT is reached after 10^{11} load cycles at the earliest. However, the hysteresis data suggests that no endurance limit exists.

9 Outlook

The results from the presented data analysis for the estimation of the different fatigue ranges lead to the question whether these developed methods can be transferred to other composites such as continuous fiber reinforced plastics. In plies with fiber reinforcement oriented transverse to the load direction, matrix cracking and failure of the fiber-matrix interface dominate the damage behavior, resulting in stiffness degradation under fatigue loading. Therefore, it can be assumed that, especially for multidirectional laminates, an estimation of the fatigue ranges can be achieved using a stiffness degradation analysis. Furthermore, with regard to the results of the fatigue data, the question arises whether a functional relationship between the investigated parameters and the fiber orientation can be derived for SFRT. In this way, analogous to the approach of the Master S-N curves [94], investigations of individual fiber orientations could allow conclusions to be drawn about other fiber orientation distributions.

The presented results of the AE analysis raise the question of whether there are signal features that allow assigning the different acoustic events to specific damage mechanisms. This would be one way to investigate which damage mechanisms are prevalent in the different fatigue ranges. In addition, further research regarding the residual strength tests with AE to discuss the existence of an endurance limit should address the transferability to other material groups.

Since virtual testing has shown to be a helpful tool, to gain insight in the micromechanical mechanisms, future investigations are to explore the incorporation of a fiber-matrix interface to investigate the damage behavior at low strains to be able to draw conclusions on the range of VHCF. For this purpose, it is necessary to further focus on the experimental characterization of the fiber-matrix interface.

Appendices

A.1: Injection molding parameters

Table A.1 Injection molding parameters

Parameter	Value	Unit
Closing pressure	550	kN
Temperatures		
Feed zone	70	°C
Heating zone 1	280	°C
Heating zone 2	285	°C
Heating zone 3	295	°C
Heating zone 4	305	°C
Nozzle	302	°C
Mold	100	°C
Feeding		
Back pressure	30	bar
Feeding stop	37	mm
Screw speed	15	m/min
Decompression	4	mm
Changeover point	8	mm
Injection		
Speed	15	mm/s
Pressure	300	bar
Hold pressure		
Hold pressure 1	650	bar
Hold pressure time 1	10	s
Hold pressure 2	600	bar
Time		
Injection time	2.4	s
Feeding time	3.3	s
Cooling time	25	s
Cycle time	42.1	s

A.2: Stress distribution in the optimized specimen geometry

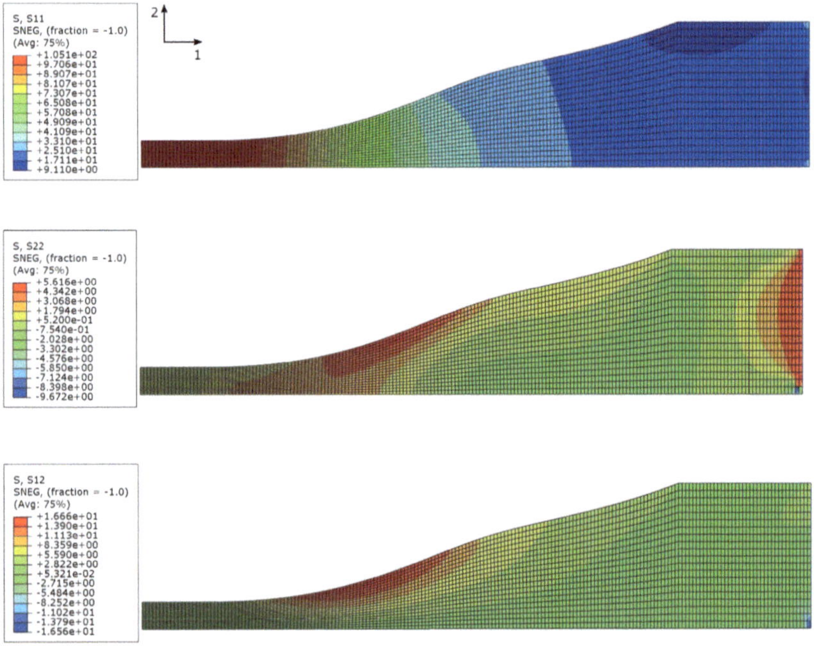

Figure A.1: Stress distributions σ_1, σ_2, and τ_{12}.

A.3: Speed of sound

The sound velocity is determined on injection molded plates for different orientations to the injection molding direction. In order to be able to position the sensors precisely, a test setup is designed and manufactured using 3D printing (Figure A.2). The speed of sound is determined using two different methods: pencil lead breakage according to Hsu-Nielson [56] and by using three sensors, one of which emits a pulse. Table A.2 lists the determined mean velocities and standard deviations.

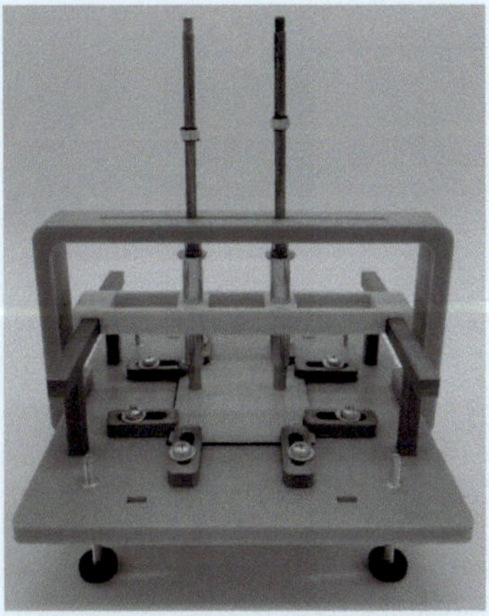

Figure A.2: Test setup for determining the speed of sound.

Table A.2 Sound velocity depending on the angle to the injection molding direction [104]

Angle to the injection molding direction	Hsu-Nielson [56]		Three sensors with pulsing function	
	Velocity in m/s	Standard deviation in m/s	Velocity in m/s	Standard deviation in m/s
0°	3274	125	3260	116
30°	3132	72	3010	85
45°	2915	51	2824	57
60°	2907	59	2812	51
90°	2759	112	2833	83

A.4: Residual strength tests for preloading with different number of load cycles

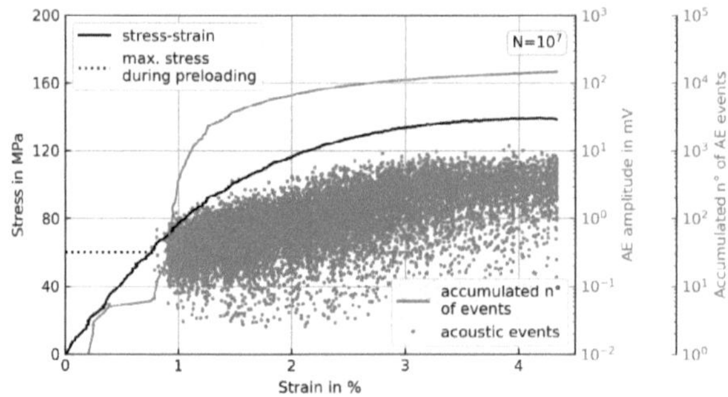

(a)

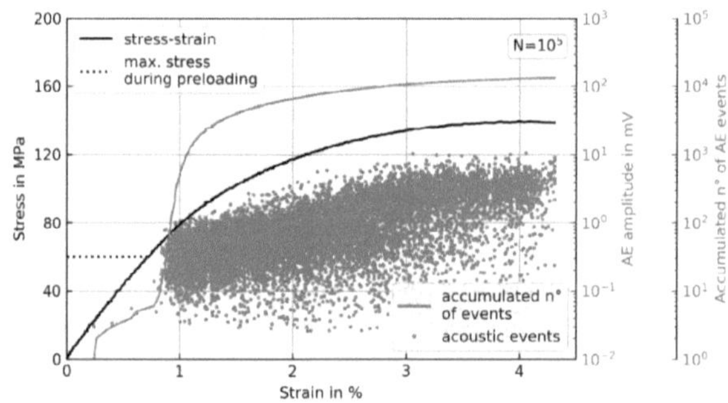

(b)

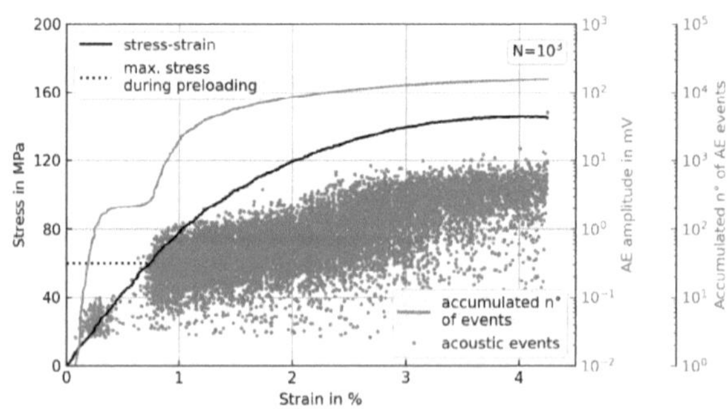

(c)

Figure A.3: Test data of 0°-specimens preloaded with σ_{max} = 40% RUTS: (a) 10^7; (b) 10^5; (c) 10^3 load cycles.

A.5: Residual strength tests of 90°-specimens

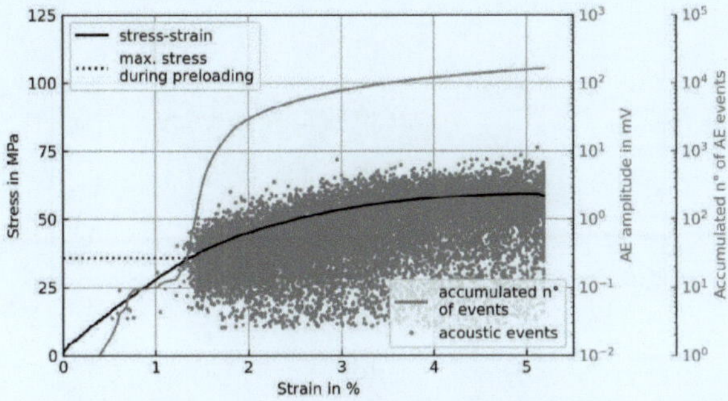

(a)

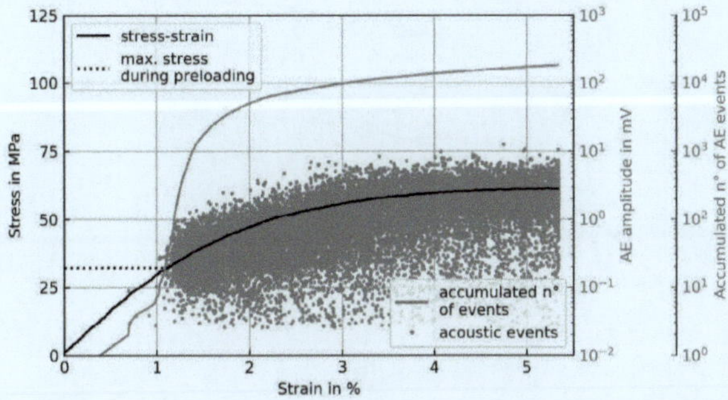

(b)

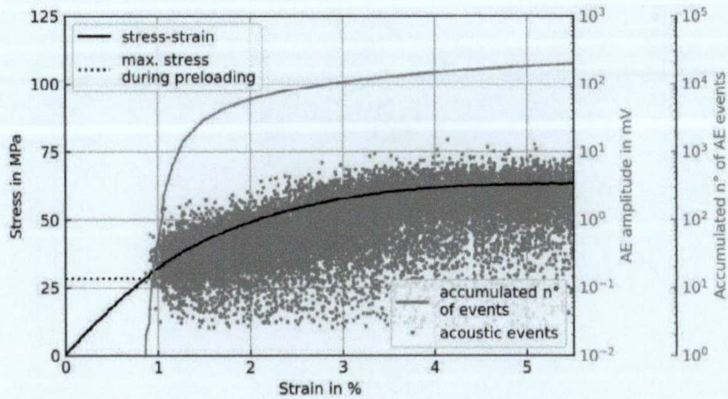

(c)

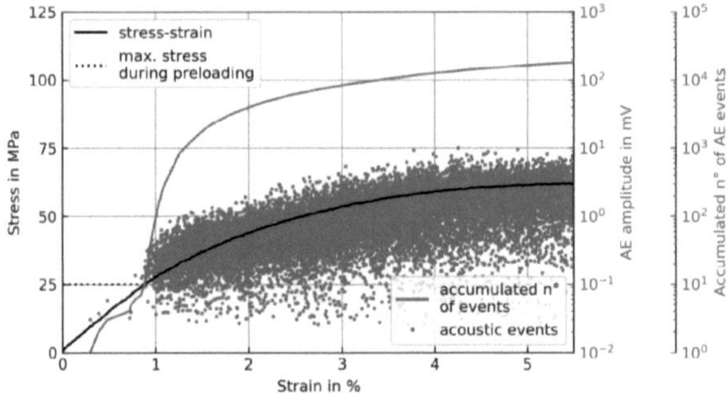
stress-strain
max. stress during preloading
accumulated n° of events
acoustic events
Stress in MPa
Strain in %
AE amplitude in mV
Accumulated n° of AE events

(d)

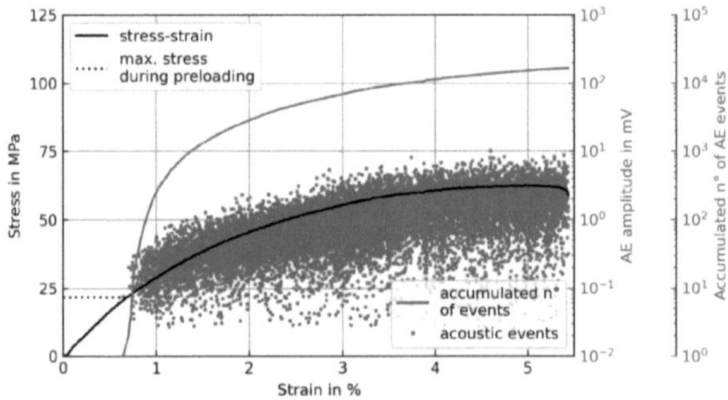
stress-strain
max. stress during preloading
accumulated n° of events
acoustic events
Stress in MPa
Strain in %
AE amplitude in mV
Accumulated n° of AE events

(e)

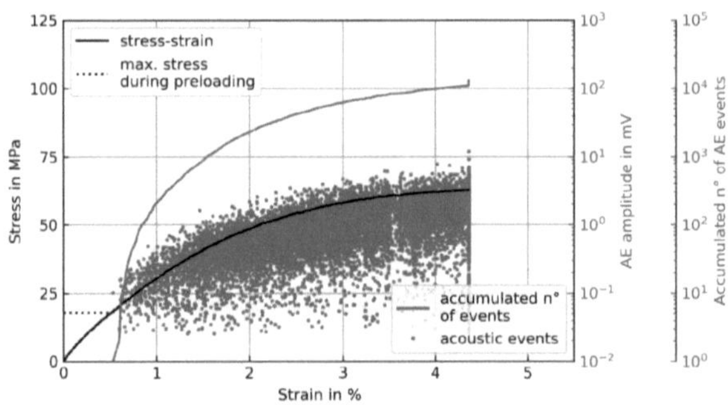
stress-strain
max. stress during preloading
accumulated n° of events
acoustic events
Stress in MPa
Strain in %
AE amplitude in mV
Accumulated n° of AE events

(f)

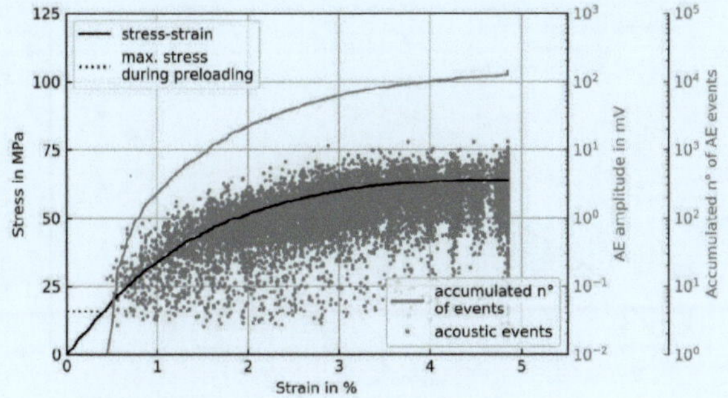
stress-strain
max. stress during preloading
accumulated n° of events
acoustic events
Stress in MPa
Strain in %
AE amplitude in mV
Accumulated n° of AE events

(g)

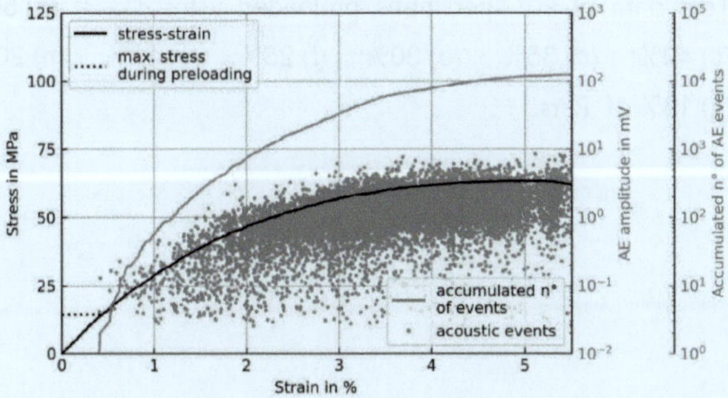
stress-strain
max. stress during preloading
accumulated n° of events
acoustic events
Stress in MPa
Strain in %
AE amplitude in mV
Accumulated n° of AE events

(h)

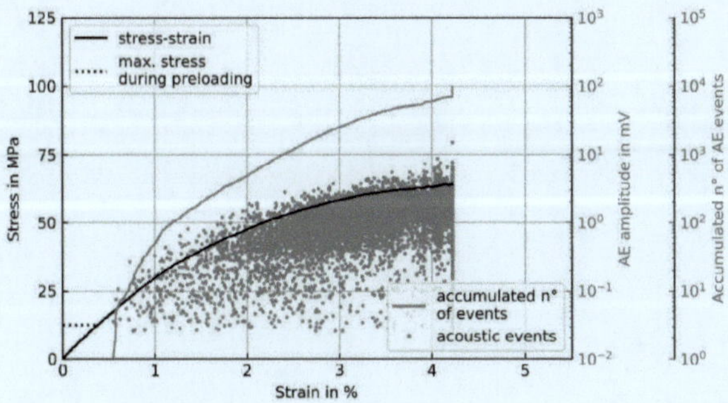
stress-strain
max. stress during preloading
accumulated n° of events
acoustic events
Stress in MPa
Strain in %
AE amplitude in mV
Accumulated n° of AE events

(i)

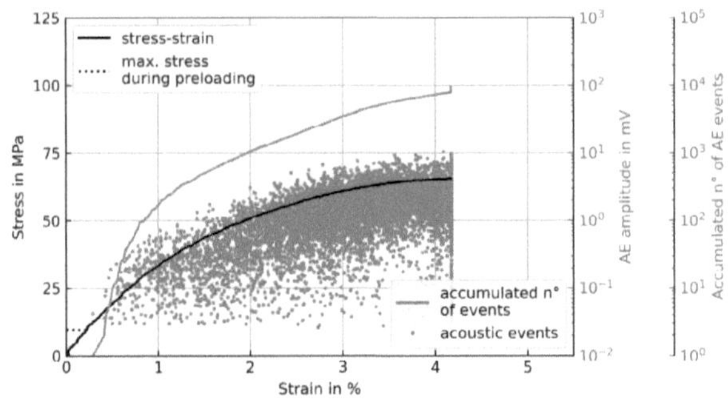

(k)

Figure A.4: Test data of 90°-specimens preloaded with: σ_{max} = (a) 50%; (b) 45%; (c) 40%; (d) 35%; (e) 30%; (f) 25%; (g) 22%; (h) 20%; (i) 18%; (k) 13% of R_{UTS}.

References

[1] Talreja, R.: Fatigue of Composite Materials. in Modern Trends in Composite Laminates Mechanics, H. Altenbach and W. Becker, Eds., pp. 281–294, Springer Vienna, Vienna, 2003.

[2] Hoppel, C. P. R., Pangborn, R. N., and Thomson, R. W.: Damage Accumulation During Multiple Stress Level Fatigue of Short-Glass-Fiber Reinforced Styrene-Maleic Anhydride. Journal of Thermoplastic Composite Materials, vol. 14, no. 1, pp. 84–94, 2001.

[3] Monte, M. de, Moosbrugger, E., and Quaresimin, M.: Influence of temperature and thickness on the off-axis behaviour of short glass fibre reinforced polyamide 6.6 – Quasi-static loading. Composites Part A: Applied Science and Manufacturing, vol. 41, no. 7, pp. 859–871, 2010.

[4] Meneghetti, G., Ricotta, M., Lucchetta, G. et al.: An hysteresis energy-based synthesis of fully reversed axial fatigue behaviour of different polypropylene composites. Composites Part B: Engineering, vol. 65, pp. 17–25, 2014.

[5] Monte, M. de, Quaresimin, M., and Lazzarin, P.: Modelling of fatigue strength data for a short fibre reinforced polyamide 6.6 based on local strain energy density. in Proceedings of 16th international conference on composite materials, Kyoto, Japan, 2007.

[6] Nouri, H., Meraghni, F., and Lory, P.: Fatigue damage model for injection-molded short glass fibre reinforced thermoplastics. International Journal of Fatigue, vol. 31, no. 5, pp. 934–942, 2009.

[7] Miller, R. K., Hill, E., and Moore, P.: Nondestructive Testing Handbook Acoustic Emission Testing. vol. 6. Columbus, OH: American Society for Nondestructive Testing, pp. 1–25, 2005.

[8] M. G. R. Sause: In situ monitoring of fiber-reinforced composites: Theory, basic concepts, methods, and applications, Springer, [Cham], op. 2016.

[9] Math2Market GmbH: GeoDict, Kaiserslautern.

[10] Brunbauer, J., Mösenbacher, A., Guster, C. et al.: Fundamental influences on quasistatic and cyclic material behavior of short glass fiber reinforced polyamide illustrated on microscopic scale. Journal of Applied Polymer Science, vol. 131, no. 19, 2014.

[11] Jia, N. and Kagan, V.: Effects of time and temperature on the tension-tension fatigue behavior of short fiber reinforced polyamides. Polymer composites, vol. 19, no. 4, pp. 408–414, 1998.

[12] Arif, M. F., Saintier, N., Meraghni, F. et al.: Multiscale fatigue damage characterization in short glass fiber reinforced polyamide-66. Composites Part B: Engineering, vol. 61, pp. 55–65, 2014.

[13] Monte, M. de, Moosbrugger, E., and Quaresimin, M.: Influence of temperature and thickness on the off-axis behaviour of short glass fibre reinforced polyamide 6.6 – cyclic loading. Composites Part A: Applied Science and Manufacturing, vol. 41, no. 10, pp. 1368–1379, 2010.

[14] Bernasconi, A., Cosmi, F., and Taylor, D.: Analysis of the fatigue properties of different specimens of a 10% by weight short glass fibre reinforced polyamide 6.6. Polymer Testing, vol. 40, pp. 149–155, 2014.

[15] Advani, S. G. and Tucker, C. L.: The Use of Tensors to Describe and Predict Fiber Orientation in Short Fiber Composites. Journal of Rheology, vol. 31, no. 8, pp. 751–784, 1987.

[16] Bernasconi, A., Conrado, E., and Hine, P.: An experimental investigation of the combined influence of notch size and fibre orientation on the fatigue strength of a short glass fibre reinforced polyamide 6. Polymer Testing, vol. 47, pp. 12–21, 2015.

[17] Bernasconi, A., Davoli, P., Basile, A. et al.: Effect of fibre orientation on the fatigue behaviour of a short glass fibre reinforced polyamide-6. International Journal of Fatigue, vol. 29, no. 2, pp. 199–208, 2007.

[18] Tanaka, K., Kitano, T., and Egami, N.: Effect of fiber orientation on fatigue crack propagation in short-fiber reinforced plastics. Engineering Fracture Mechanics, vol. 123, pp. 44–58, 2014.

[19] Zhou, Y. and Mallick, P. K.: Fatigue performance of an injection-molded short E-glass fiber-reinforced polyamide 6,6. I. Effects of orientation, holes, and weld line. Polymer Composites, vol. 27, no. 2, pp. 230–237, 2006.

[20] Horst, J. J. and Spoormaker, J. L.: Mechanisms of fatigue in short glass fiber reinforced polyamide 6. Polymer Engineering & Science, vol. 36, no. 22, pp. 2718–2726, 1996.

[21] Klimkeit, B., Castagnet, S., Nadot, Y. et al.: Fatigue damage mechanisms in short fiber reinforced PBT+PET GF30. Materials Science and Engineering: A, vol. 528, no. 3, pp. 1577–1588, 2011.

[22] Belmonte, E., De Monte, M., Quaresimin, M. et al.: Damage Mechanisms of Short Glass Fiber Reinforced Polyamide 6.6 under Fatigue Loading. in 16th European Conference on Composite Materials (ECCM 2014), 2014.

[23] Rolland, H., Saintier, N., and Robert, G.: Damage mechanisms in short glass fibre reinforced thermoplastic during in situ microtomography tensile tests. Composites Part B: Engineering, vol. 90, pp. 365–377, 2016.

[24] ISO, ISO 13003: Fibre-reinforced plastics - Determination of fatigue properties under cyclic loading conditions.

[25] Harris, B.: Fatigue in composites: science and technology of the fatigue response of fibre-reinforced plastics, Woodhead Publishing, 2003.

[26] Eftekhari, M. and Fatemi, A.: On the strengthening effect of increasing cycling frequency on fatigue behavior of some polymers and their composites: Experiments and modeling. International Journal of Fatigue, vol. 87, pp. 153–166, 2016.

[27] Bellenger, V., Tcharkhtchi, A., and Castaing, P.: Thermal and mechanical fatigue of a PA66/glass fibers composite material. International Journal of Fatigue, vol. 28, no. 10, pp. 1348–1352, 2006.

[28] Handa, K., Atsushi, K., and Narisawa, I.: Fatigue characteristics of a glass-fiber-reinforced polyamide. Journal of Applied Polymer Science, no. 72, 1999.

[29] Launay, A., Maitournam, M. H., Marco, Y. et al.: Cyclic behaviour of short glass fibre reinforced polyamide: Experimental study and constitutive equations. International Journal of Plasticity, vol. 27, no. 8, pp. 1267–1293, 2011.

[30] Meneghetti, G., Ricotta, M., Colombera, G. et al.: The influence of reinforcement on fatigue behaviour of a polypropylene composite. in ECCM15—15th European Conference on Composite Materials, 2012.

[31] Sonsino, C.M. and Moosbrugger, E.: Fatigue design of highly loaded short-glass-fibre reinforced polyamide parts in engine compartments. International Journal of Fatigue, vol. 30, no. 7, pp. 1279–1288, 2008.

[32] Cosmi, F. and Bernasconi, A.: Micro-CT investigation on fatigue damage evolution in short fibre reinforced polymers. Composites Science and Technology, vol. 79, pp. 70–76, 2013.

[33] Czochralski, J.: Metallographische Untersuchungen am Zinn und ihre fundamentale Bedeutung fuer die Theorie der Formaenderung bildsamer Metalle. Int Zeitschrift fuer Metallographie, vol. 8, pp. 1–43, 1916.

[34] Kaiser, J.: Untersuchungen über das Auftreten von Geräuschen beim Zugversuch, Technische Hochschule München, 1950.

[35] Godin, N., Reynaud, P., and Fantozzi, G.: Acoustic emission and durability of composite materials, John Wiley & Sons, 2018.

[36] Ohtsu, M.: The history and development of acoustic emission in concrete engineering. Magazine of concrete research, vol. 48, no. 177, pp. 321–330, 1996.

[37] Gholizadeh, S., Leman, Z., and Baharudin, B. H. T.: A review of the application of acoustic emission technique in engineering. Structural Engineering and Mechanics, vol. 54, no. 6, pp. 1075–1095, 2015.

[38] Shevchik, S. A., Kenel, C., Leinenbach, C. et al.: Acoustic emission for in situ quality monitoring in additive manufacturing using spectral convolutional neural networks. Additive Manufacturing, vol. 21, pp. 598–604, 2018.

[39] Grosse, C. U., Ohtsu, M., Aggelis, D. G. et al.: Acoustic Emission Testing: Basics for Research - Applications in Engineering, Springer International Publishing, Cham, 2022.

[40] Nordstrom, R. A.: Acoustic emission characterization of microstructural failure in the single fiber fragmentation test, EMPA, 1996.

[41] Sause, M.G.R.: Identification of failure mechanisms in hybrid materials utilizing pattern recognition techniques applied to acoustic emission signals, Dissertation, Universität Augsburg, Dezember 2010.

[42] Ono, K.: Through-transmission characteristics of AE sensor couplants. Journal of Acoustic Emission, vol. 34, pp. 1–11, 2017.

[43] Richler, S.: Gezielte Erzeugung mikroskopischer Schädigungsarten von Faserverbundwerkstoffen in Kombination mit Schallemissionsanalyse, 2015.

[44] Scruby, C. B.: An introduction to acoustic emission. Journal of Physics E: Scientific Instruments, vol. 20, no. 8, p. 946, 1987.

[45] Mehan, R. L. and Mullin, J. V.: Analysis of composite failure mechanisms using acoustic emissions. Journal of composite materials, vol. 5, no. 2, pp. 266–269, 1971.

[46] Sachse, W. and Kim, K. Y.: Quantitative acoustic emission and failure mechanics of composite materials. Ultrasonics, vol. 25, no. 4, pp. 195–203, 1987.

[47] Kurz, J. H., Grosse, C. U., and Reinhardt, H.-W.: Strategies for reliable automatic onset time picking of acoustic emissions and of ultrasound signals in concrete. Ultrasonics, vol. 43, no. 7, pp. 538–546, 2005.

[48] Kelkel, B.: On the influence of source depth and source-to-sensor distance on the acoustic emission signatures of damaging events in cross-ply carbon fibre-reinforced plastics, Technische Universität Kaiserslautern, 2020.

[49] Akaike, H.: Markovian Representation of Stochastic Processes and Its Application to the Analysis of Autoregressive Moving Average Processes, 1974.

[50] Zhang, H., Thurber, C., and Rowe, C.: Automatic P-wave arrival detection and picking with multiscale wavelet analysis for single-component recordings. Bulletin of the Seismological Society of America, vol. 93, no. 5, pp. 1904–1912, 2003.

[51] Maeda, N.: A method for reading and checking phase times in autoprocessing system of seismic wave data. Zisin, vol. 38, pp. 365–379, 1985.

[52] Graff, K. F.: Wave motion in elastic solids, Clarendon Press, Oxford, 1975.

[53] Achenbach, J. D.: Wave Propagation in elastic solids, North-Holland, Amsterdam, 1980.

[54] Lamb, H.: On waves in an elastic plate. Proceedings of the Royal Society of London. Series A, Containing papers of a mathematical and physical character, vol. 93, no. 648, pp. 114–128, 1917.

[55] Putkis, O., Dalton, R. P., and Croxford, A. J.: The anisotropic propagation of ultrasonic guided waves in composite materials and implications for practical applications. Ultrasonics, vol. 65, pp. 390–399, 2016.

[56] Hsu, N. N. and Breckenridge, F. R.: Characterization and calibration of acoustic emission sensors. Materials Evaluation, vol. 39, pp. 60–68, 1981.

[57] Barré, S. and Benzeggagh, M. L.: On the use of acoustic emission to investigate damage mechanisms in glass-fibre-reinforced polypropylene. Composites Science and Technology, vol. 52, no. 3, pp. 369–376, 1994.

[58] Berthelot, J. M. and Rhazi, J.: Different Types of Amplitude Distributions in Composite Materials. Journal of Reinforced Plastics and Composites, vol. 7, no. 4, pp. 302–320, 1988.

[59] Bohse, J.: Acoustic emission characteristics of micro-failure processes in polymer blends and composites. Composites Science and Technology, vol. 60, no. 8, pp. 1213–1226, 2000.

[60] Sato, N., Kurauchi, T., Sato, S. et al.: Mechanism of fracture of short glass fibre-reinforced polyamide thermoplastic. Journal of materials science, vol. 19, no. 4, pp. 1145–1152, 1984.

[61] Arumugam, V., Kumar, C. S., Santulli, C. et al.: Identification of Failure Modes in Composites from Clustered Acoustic Emission Data Using Pattern Recognition and Wavelet Transformation. Arabian Journal for Science and Engineering, vol. 38, no. 5, pp. 1087–1102, 2013.

[62] Gutkin, R., Green, C. J., Vangrattanachai, S. et al.: On acoustic emission for failure investigation in CFRP: Pattern recognition and peak frequency analyses. Mechanical Systems and Signal Processing, vol. 25, no. 4, pp. 1393–1407, 2011.

[63] Ramirez-Jimenez, C.R., Papadakis, N., Reynolds, N. et al.: Identification of failure modes in glass/polypropylene composites by means of the primary frequency content of the acoustic emission event. Composites Science and Technology, vol. 64, no. 12, pp. 1819–1827, 2004.

[64] Carvelli, V., D'Ettorre, A., and Lomov, S. V.: Acoustic emission and damage mode correlation in textile reinforced PPS composites. Composite Structures, vol. 163, pp. 399–409, 2017.

[65] Oz, F. E., Ersoy, N., Mehdikhani, M. et al.: Multi-instrument in-situ damage monitoring in quasi-isotropic CFRP laminates under tension. Composite Structures, vol. 196, pp. 163–180, 2018.

[66] Tang, J., Soua, S., Mares, C. et al.: A Pattern Recognition Approach to Acoustic Emission Data Originating from Fatigue of Wind Turbine Blades. Sensors (Basel, Switzerland), vol. 17, no. 11, 2017.

[67] Schoßig, M., Zankel, A., Bierögel, C. et al.: ESEM investigations for assessment of damage kinetics of short glass fibre reinforced thermoplastics – Results of in situ tensile tests coupled with acoustic emission analysis. Composites Science and Technology, vol. 71, no. 3, pp. 257–265, 2011.

[68] Benz, J., Willems, F., and Bonten, C.: The Sound of Materials. Kunststoffe international, no. 11, pp. 46–49, 2016.

[69] Bauer, C., Rief, T., Hausmann, J. et al.: A comparative study of the static and cyclic acoustic emission behavior of fiber reinforced thermoplastics. in 17th European Conference on Composite Materials (ECCM17), 2016.

[70] Rauter, N. and Lammering, R.: Experimental Characterization of Short Fiber-Reinforced Composites on the Mesoscale by Indentation Tests. Applied Composite Materials, vol. 28, no. 5, pp. 1747–1765, 2021.

[71] Rui Melro, A.: Analytical and numerical modelling of damage and fracture of advanced composites, Universidade do Porto.

[72] Zohdi, T. I. and Wriggers, P.: An introduction to computational micromechanics, Springer Science & Business Media, 2004.

[73] Eshelby, J. D.: The determination of the elastic field of an ellipsoidal inclusion, and related problems. in Proceedings of the Royal Society of London A: Mathematical, Physical and Engineering Sciences, vol. 241, pp. 376–396, 1957.

[74] Mori, T. and Tanaka, K.: Average stress in matrix and average elastic energy of materials with misfitting inclusions. Acta metallurgica, vol. 21, no. 5, pp. 571–574, 1973.

[75] Halpin, J. C.: Stiffness and expansion estimates for oriented short fiber composites. Journal of composite materials, vol. 3, no. 4, pp. 732–734, 1969.

[76] Tucker III, C. L. and Liang, E.: Stiffness predictions for unidirectional short-fiber composites: Review and evaluation. Composites Science and Technology, vol. 59, no. 5, pp. 655–671, 1999.

[77] Giannaros, E., Kotzakolios, T., Tsantzalis, S. et al.: Implementation and calibration of meso-scale modeling technique for simulation of tensile behavior of fabric materials. Composites Part B: Engineering, vol. 119, pp. 1–9, 2017.

[78] Zhang, C., Binienda, W. K., Goldberg, R. K. et al.: Meso-scale failure modeling of single layer triaxial braided composite using finite element method. Composites Part A: Applied Science and Manufacturing, vol. 58, pp. 36–46, 2014.

[79] Krairi, A. and Doghri, I.: Multi-scale Damage Model for Mechanical High Cycle Fatigue (HCF) of Short Glass Fibre Reinforced Thermoplastics (SGFRTP). Procedia Engineering, vol. 66, pp. 759–765, 2013.

[80] Wilmes, A. and Hornberger, K.: Influence of Fiber Orientation and Multiaxiality on the Fatigue Strength of Unnotched Specimens – Lifetime Estimation. Procedia Engineering, vol. 133, pp. 148–160, 2015.

[81] Hill, R.: Elastic properties of reinforced solids: Some theoretical principles. Journal of the Mechanics and Physics of Solids, vol. 11, no. 5, pp. 357–372, 1963.

[82] Drugan, W. J. and Willis, J. R.: A micromechanics-based nonlocal constitutive equation and estimates of representative volume element size for elastic composites. Journal of the Mechanics and Physics of Solids, vol. 44, no. 4, pp. 497–524, 1996.

[83] Breuer, K. and Stommel, M.: RVE modelling of short fiber reinforced thermoplastics with discrete fiber orientation and fiber length distribution. SN Applied Sciences, vol. 2, no. 1, 2020.

[84] Huang, H.-B. and Huang, Z.-M.: Micromechanical prediction of elastic-plastic behavior of a short fiber or particle reinforced composite. Composites Part A: Applied Science and Manufacturing, vol. 134, p. 105889, 2020.

[85] Fliegener, S., Kennerknecht, T., and Kabel, M.: Simulation of the damage mechanisms of glass fiber reinforced polypropylen based on micro specimens and 1:1 models of their microstructure. in ECCM17: 17th European Conference on Composite Materials, Munich, Germany, 26 - 30th June 2016, 2016.

[86] Li, Z., Liu, Z., Lei, Z. et al.: An innovative computational framework for the analysis of complex mechanical behaviors of short fiber reinforced polymer composites. Composite Structures, vol. 277, p. 114594, 2021.

[87] Fliegener, S., Kennerknecht, T., and Kabel, M.: Investigations into the damage mechanisms of glass fiber reinforced polypropylene based on micro specimens and precise models of their microstructure. Composites Part B: Engineering, vol. 112, pp. 327–343, 2017.

[88] Moulinec, H. and Suquet, P.: A fast numerical method for computing the linear and nonlinear mechanical properties of composites. Comptes rendus de l'Académie des sciences. Série II. Mécanique, physique, chimie, astronomie, 1994.

[89] Kröner, E.: Statistical continuum mechanics, Springer, 1972.

[90] Kabel, M., Fink, A., and Schneider, M.: The composite voxel technique for inelastic problems. Computer Methods in Applied Mechanics and Engineering, vol. 322, pp. 396–418, 2017.

[91] Kabel, M., Merkert, D., and Schneider, M.: Use of composite voxels in FFT-based homogenization. Computer Methods in Applied Mechanics and Engineering, vol. 294, pp. 168–188, 2015.

[92] Spahn, J., Andrä, H., Kabel, M. et al.: A multiscale approach for modeling progressive damage of composite materials using fast Fourier transforms. Computer Methods in Applied Mechanics and Engineering, vol. 268, pp. 871–883, 2014.

[93] Bauer, C.: Charakterisierung und numerische Beschreibung des nichtlinearen Werkstoff- und Lebensdauerverhaltens eines kurzglasfaserverstärkten Polymerwerkstoffes unter Berücksichtigung der im µCT gemessenen lokalen Faserorientierung, Dissertation, TU Kaiserslautern, 2017.

[94] Carvelli, V., Jain, A., Lomov, S.: Fatigue of Textile and Short Fibre Reinforced Composites, ISTE Ltd and John Wiley & Sons, Inc., London, 2017.

[95] Mortazavian, S. and Fatemi, A.: Tensile behavior and modeling of short fiber-reinforced polymer composites including temperature and strain rate effects. Journal of Thermoplastic Composite Materials, vol. 30, no. 10, pp. 1414–1437, 2017.

[96] Zhou, Y. and Mallick, P. K.: A non-linear damage model for the tensile behavior of an injection molded short E-glass fiber reinforced polyamide-6,6. Materials Science and Engineering: A, vol. 393, 1-2, pp. 303–309, 2005.

[97] DIN Deutsches Institut für Normung e.V.: DIN EN ISO 527-1: Kunststoffe-Bestimmung der Zugeigenschaften-Teil 1: Allgemeine Grundsätze.

[98] Fowler, T. J.: Acoustic Emission Testing of Fiber Reinforced Plastics, San Francisco, California, Oct. 17-21, 1977.

[99] Gloeckner, R., Kolling, S., and Heiliger, C.: A Monte-Carlo Algorithm for 3D Fibre Detection from Microcomputer Tomography. Journal of Computational Engineering, vol. 2016, pp. 1–9, 2016.

[100] Thomason, J. L.: Structure–property relationships in glass reinforced polyamide, part 2: The effects of average fiber diameter and diameter distribution. Polymer Composites, vol. 28, no. 3, pp. 331–343, 2007.

Student thesises

[101] Stubbe, L.: Validierung von kommerzieller FE-Software zur Lebensdauerabschätzung von Faser-Kunststoff-Verbundwerkstoffen. Bachelorarbeit, Kaiserslautern, IVW-Bericht, 2017, 17-087.

[102] Rainot, G.: Vergleich von FE-Software zur Lebensdauerabschätzung von Faser-Kunststoff-Verbundwerkstoffen. Studienarbeit, Kaiserslautern, IVW-Bericht, 2018, 18-001.

[103] Schmitt, F.: Entwicklung einer Prüfmethodik zur Lebensdauerbewertung von Schwungrädern zum Einsatz als Energiespeicher, Masterarbeit, Kaiserslautern, IVW-Bericht, 2018, 18-023

[104] Detzel, M.: Untersuchung des Einflusses der Faserorientierung auf das akustische Signal von kurzfaserverstärkten Thermoplasten, Bachelorarbeit, Kaiserslautern, IVW-Bericht, 2021, 21-006

Conferences and publications

[105] Krummenacker, J., Klingler, A., Hausmann, J., Wetzel, B.: Improvement of the cyclic loading resistance of highly loaded CFRP components by resin modifications. In: Technical Report TX1RP201720694, Airbus Group Innovations, 2017.

[106] Krummenacker, J., Hausmann, J., Sorochynska, L., Klingler, A., Wetzel, B.: Development of a cyclic test method for ultra-fast rotating flywheels made of CFRP and improvement of their fatigue strength by matrix modifications. In: 21st International Conference on Composite Materials (ICCM21). Xi'An, China, 2017.

[107] Krummenacker, J.: Lastwechselfeste Harze für Energiespeicher-Anwendungen. In Carbon Composites Magazin, 1/2018.

[108] Hausmann, J., Krummenacker, J.: Ganzheitliche Bauteilentwicklung am Beispiel ermüdungsbelasteter Strukturen, 23. Nationales SAMPE Symposium, Kaiserslautern, Germany, 28.02. – 01.03.2018.

[109] Krummenacker, J.; Hausmann, J.; Klingler, A.; Wetzel, B.: Life-time prolongation by matrix modifications of flywheel components made of carbon fiber reinforced polymers. 7th International Conference on Fatigue of Composites, Vicenza, Italy, 04.-06.07.2018.

[110] Hausmann, J., Krummenacker, J., Klingler, A., Wetzel, B.: Improvement of fatigue strength of carbon fiber reinforced polymers by matrix modifications for ultra-fast rotating flywheels, in "Frontiers in Science and Technology - Reports on technologies for sustainability" edited by O. Kanoun, 2021.

[111] Krummenacker, J.; Hausmann, J.: Determination of a Threshold for Initiation of Damage in Short Glass Fiber Reinforced Polyamide 66, 8th International Conference on Fatigue of Composites, Virtual Conference, 23. – 25.06.2021.

[112] Krummenacker, J.; Hausmann, J.: Determination of Fatigue Damage Initiation in Short Fiber-Reinforced Thermoplastic through Acoustic Emission Analysis. J. Compos. Sci. 2021, 5, 221.

[113] Hausmann, J.; Esha, E.; Schmidt, S.; Krummenacker, J.: Transfer of mechanical properties from specimens to injection molded structures under consideration of local fiber orientation. EUROMAT 2021 - Virtual Conference, 12. – 17.09.2021.

List of figures

List of tables

Curriculum Vitae

Personal information

	Janna Krummenacker
	Born in Norden, Germany
	German

Secondary education

09/1996 – 03/2005	High-School, Albert-Schweitzer-Gymnasium, Kaiserslautern (Germany)
08/2002 – 02/2003	Lake Oswego High School, Lake Oswego (United States of America): Semester abroad

Academic education

10/2005 – 04/2011	Universität Karlsruhe (TH), later Karlsruher Institut für Technologie, Karlsruhe (Germany): Diploma, Civil engineering
08/2007 – 07/2008	Institut National des Sciences Appliquées, Lyon (France): Semester abroad
11/2010 – 02/2011	École Polytechnique Fédérale, Lausanne (Switzerland): Diploma thesis

Work experience

09/2011 – 12/2014	Leonhardt, Andrä und Partner AG, Stuttgart (Germany): Structural engineer
01/2015 – 07/2015	Fachgebiet Massivbau und Baukonstruktion, Technische Universität Kaiserslautern (Germany): Research Assistant
08/2015 – present	Leibniz-Institut für Verbundwerkstoffe GmbH, Kaiserslautern (Germany): Research assistant